THE BEE KEEPER'S FIELD GUIDE

THE BEE KEEPER'S FIELD GUIDE

A pocket guide to the health and care of bees

DAVID CRAMP

A How To Book

ROBINSON

First published in Great Britain in 2011 by Spring Hill, an imprint of How To Books Ltd
This edition published in Great Britain in 2016 by Robinson

1 3 5 7 9 10 8 6 4 2

NOTE: The material contained is for general guidance and does not deal with particular circumstances. Laws and regulations are complex and liable to change, and readers should check the current position with relevant authorities or consult appropriate financial advisers before making personal arrangements.

A CIP catalogue record for this book is available from the British Library.

ISBN 978-1-4721-3847-7

Typeset by Baseline Arts Ltd
Printed and bound in Great Britain by Bell & Bain Ltd

Papers used by Robinson are from well-managed forests and other responsible sources

Robinson is an imprint of Little, Brown Book Group
Carmelite House
50 Victoria Embankment
London EC4Y 0DZ

An Hachette UK Company
www.hachette.co.uk

www.littlebrown.co.uk

How To Books are published by Robinson, an imprint of Little, Brown Book Group. We welcome proposals from authors who have first-hand experience of their subjects. Please set out the aims of your book, its target market and its suggested contents in an email to Nikki.Read@howtobooks.co.uk

Contents

Part G: Hive Checks

Part H: Pests and Diseases

Part I: A Beekeeper's Ready-reckoner

Part J

Preface

The object of this book is not to teach beekeeping but to provide a handy beekeeping guide that can be taken into and used in the apiary, as an operational, troubleshooting and disease-diagnosis reference. For a beekeeper, there is so much to know that, out in the field away from manuals and reference books and usually alone, it is often difficult to remember the exact sequence of events for certain tasks and manipulations, or the precise symptoms of a particular disease.

This guide is an attempt to overcome that. It is meant for the beekeeper to use in the field as an aide-memoire and diagnostic tool – something to jog the memory in a variety of situations. Before you leave the apiary, take a glance at the appropriate reference in the book just to make sure that you have remembered to complete everything and that you have picked up your hive tool.

This guide is aimed primarily at the hobbyist beekeeper who knows what they want to do but may, at times, need a hand at doing it. However, beekeepers of all levels will find it useful, and it may suggest a number of alternative methods to professional beekeepers.

I wrote the guide initially for myself because I kept bees in fairly remote areas and often needed to refer to one manual or another over something while out in the field. Kept in a locker in the car, these excellent but usually very expensive books would end up torn and filthy in no time, with half the pages coming adrift. Keep this one in your bee bag, take it into the field and use it.

Figures

Photographs

Tables

Picture credits

All the photos are copyright of the author except the following which are gratefully acknowledged:

Photo: 2. Colin Eastham
Photos: 3,4,14,19,20,21,22,23,32,33,34,35,36,39,40. The Food and Environment Research Agency (FERA), reproduced under the terms of the Click-Use Licence. No: C2010002634
Photos: 9,18,24,25,31. International Bee Research Association (IBRA)
Photo 16. D Hobern, licensed under the Creative Commons Attribution 2.0 Generic License.

Introduction

There are many variations on a theme, and beekeeping is no exception to this. As this is a field briefing guide, only those manipulations that offer a fair chance of success, based on personal experience and the results of bee research, have been included, and only those diseases where it is possible to make a field diagnosis are listed in Part H. Acarine (tracheal mite) is included in order to impress on beekeepers that, despite popular opinion, there are few – if any – symptoms visible in the field other than a dwindling of the colony. New and exotic pests are detailed as well as the poorly understood (as yet) parasitic mite syndrome and colony collapse disorder, both of which have a definite and evident field diagnosis. It is in every beekeeper's interest to keep a keen eye out for those problems, such as Tropilaelaps clarae and the small hive beetle, which are not yet in evidence in many areas.

Part A of the guide provides checklists of the important points to consider when siting and setting up an apiary in both country and urban areas. In these days of swift and instant litigation, it is important that beekeepers know and understand their responsibilities both in regard to the welfare of their bees and consideration of their neighbours. The rules for establishing an organic honey-producing apiary are also listed for those who wish to take beekeeping one step further.

Part B of the guide supplies lists of nectar-bearing plants and crops, pollen providers and those plants that will produce honeydew honey. Planting an apiary in the middle of a million acres of wheat will benefit no one, and so knowing which plants are in your area of operations and when they are likely to give nectar is of crucial importance in running a successful beekeeping hobby or business. It is important to remember, however, that you must learn about flowers, climate and timings in your own area or an area where you plan to set up an apiary. A general list can offer only an average of all the known facts, and knowledge will come with experience.

Part C provides both a series of troubleshooting guides for problems commonly associated with colonies and a series of checklists for manipulations designed to overcome these and other problems. Without doubt, there are many more methods of achieving a desired aim than those discussed here, both 'secret' and well known, but I have carried out each of the methods listed in this field guide and, while nothing is certain with bees, these methods have a high degree of success.

Part D offers field guidance on swarm prevention and control methods as well as guidance on setting up swarm traps to increase your own stock. This aspect of beekeeping is extremely important in apiary management and deserves special study. There is no more important aspect of beekeeping than controlling swarming if you want both to enjoy beekeeping and to produce an abundance of surplus honey during the season, yet it is one that many beekeepers ignore, at worst, and just hope will go away, at best. You are fighting an almost irresistible urge – one of reproduction, albeit at colony level – but, in most cases, this can be accomplished without resorting to methods that go against the direction the bees want to go in.

Part E introduces various aspects of queen rearing and queen supersedure. As the chief production unit of the colony, carrying out the correct procedures is of great importance for achieving the aim of rearing good queens. Follow the procedures in the guide carefully to ensure the health and success of your queens.

Part F details honey harvest procedures both prior to the harvest and during, and post-harvest, and it is designed to assist beekeepers achieve their goal of an uneventful, successful harvest. Many beekeepers, especially novices, start to worry about the harvest because they are entering new territory, but in fact this is one of the easiest parts of beekeeping if all the procedures are followed. Post-harvest honey checks are included and it should be remembered that the correct storage of honey, especially if it is for sale, is important.

Part G provides a list of the all important seasonal checks that will help you ensure your colonies are moving in the right direction throughout the year. It is always a relief when in the field to be able to refer to a written guide when carrying out checks because, undoubtedly, you will remember that one check you forgot as soon as you arrive home. Keep the hives open until you've gone through the lists and all should be well. And remember to pick up your hive tool before you leave the apiary – it's always the item you lose first.

The comprehensive 'disease field-diagnosis guide' in **Part H** covers most common pests and pathogens and stresses where visual diagnosis is difficult. Varroa is included as well as the not fully understood parasitic mite syndrome and colony collapse disorder, but remember that there are new research findings every week and that beekeepers must keep up to date with these in order to cope with these problems. Also remember that, if you are in a Varroa-listed area and you haven't got Varroa, then you have got some remarkable bees. Let someone know. You don't get a vet in beekeeping so it's up to you!

Part I includes information for beekeepers out in the field in ready-reckoner form. All beekeepers get blanks when trying to remember a conversion factor or how long a drone is in the cell before emergence, or how exactly to do a quick test for Varroa mite numbers! There is a wealth of useful information here at your fingertips, including guidance on the pollination requirements of various crops which you must know if you are talking 'contracts' with a prospective client.

The references in **Part J** form a bibliography of researched information and interesting articles written by experts in their field. I have also included a list of those books I believe to be the classics on the craft and science of beekeeping. Only the main global beekeeping organisations have been included in this part. Many smaller ones tend to fade away or change form or address, and the *Beekeeper's Annual* from Northern Bee Books is a very useful and detailed source of this information. A comprehensive index concludes the guide.

Part A

Setting up an apiary

Introduction

Locating satisfactory apiary sites, whether in a rural or urban environment, is one of the most important fundamentals of successful beekeeping. Just anywhere won't do, and it's absurd to think that bees can prosper in some of the sites allocated to them by beekeepers – frost hollows, sites prone to flooding, sites open to high winds and driving rain, and so on. Bees wouldn't choose these areas and so neither should the beekeeper. One of the main roles of any farmer is to provide adequate living conditions for the farm's livestock, whether cows, sheep or bees.

Bees are hardy creatures but, if you want them to provide you with surplus honey, make sure that, right from the start, you give them well situated apiary sites with access to ample food sources and water. This fundamental responsibility is often overlooked, resulting in poor colonies prone to disease and never really prospering. Many commercial beekeepers have to place their hives wherever they can get permission from a landowner, but the general principles apply as much to them as to the hobbyist with two hives. It isn't always possible to get it spot on every time but do your best.

SECTION 1:
APIARY SITE CHECKLISTS

Various factors must be taken into account when siting hives, especially permanent sites. The following are ideal locations.

1. *Prevent drifting – place hives randomly*

2. *Placing hives on pallets in a horseshoe pattern to prevent drifting*

Rural areas

- Only site apiaries in places that have easy access on foot and in a vehicle.
- Site hives in a random pattern or in a horseshoe/circular pattern to avoid drifting (see Photographs 1 and 2). (Hives in a scattered pattern with entrances in different directions are fine, but make sure they are not so random that the hive you are inspecting isn't in the direct flight path of other hives).
- Permanent sites should have nectar sources within a 1 mile radius but preferably nearer. Ensure that nectar sources cover most of the year (see Sections 5–8), unless you are prepared to move the bees when required.
- Sites should have a nearby source of water. This should be in full sun and out of the wind. This is an important requirement.

- Sites should have a source(s) of early pollen-yielding plants or you should be prepared to feed pollen substitute early in the year (see Section 58).
- In temperate climates sites should be in full sun. Some tree shade is useful but don't site hives under trees where they can be dripped on in the rain.
- Sites should be sheltered from wind if possible. In very exposed areas place hay bales around them for protection from the wind.
- Ensure that apiary sites are not prone to flooding in winter. In extreme winters even what look to be safe sites can be flooded.
- Ensure apiary sites are not in winter frost hollows.
- Always allow ventilation underneath hives, especially if using a mesh floor. Place the hives on secure stands.
- Apiaries should preferably be out of sight of roads to avoid both theft and vandalism.
- Insure your hives for damage and theft.

Urban areas

Urban apiary sites have the same requirements as those in the country but there are additional factors to consider. The main factor is neighbours, and these should be foremost in your mind. To ensure that urban apiaries do not interfere with your neighbours, remember the following:

- Rooftops keep bees away from neighbours, and many beekeepers in cities now keep hives on their roofs (out of sight, out of mind).
- Use a gentle strain of bees. Italians or Cecropians are gentle and easy to manipulate.
- Keep swarming to a minimum (see Part D).
- Prevent robbing (see Section 13). Robbing can be alarming to neighbours who will see it as bees being out of control (which they are!).
- Try to keep the hives concealed (roofs are ideal).
- Place a fence in front of the hive entrances to keep the bees flying high.
- Have a water source nearby right at the start so that your bees don't infest garden ponds/swimming pools, etc., belonging to your neighbours.
- Provide neighbours with honey.
- Insure your bees for third-party liability (see your local beekeeping association).

SECTION 2:
OBTAINING AND PLACING BEES

Usually, bees arrive as a 'hive of bees', 'a nucleus of bees' (usual in Europe), 'a package of bees' (usual in the USA) or as a 'swarm'. This section provides a checklist of actions that you must take to ensure that the bees settle in to their new location with the minimum of disruption (see Section 1 for a checklist of ideal sites).

Placing a hive of bees

- Before their arrival, prepare a hive stand to ensure that the hive is off the ground.
- Place the newly arrived hive on the stand.
- Tilt the hive slightly forward to ensure that rain cannot enter.
- Open the hive entrance.
- Ensure that the hive is adequately strapped.
- Leave all alone for a few days and then carry out a full hive inspection (see Section 10).

Placing a nucleus (nuc) of bees

- Place each nucleus on the lid of the hive it will occupy with the entrance facing the same way as the hive entrance. Leave until the following evening.
- The next day place the nucleus to one side and open the hive.
- Carefully lift out one frame at a time from the nuc and place it in the hive, ensuring you don't lose the queen in the process.
- Exchange an outside frame from the brood box for a frame feeder with 1:1 syrup.

- Once all the frames from the nuc are in the hive with the feeder, close up the hive.
- Reduce the hive's entrance to one or two bee spaces either with a custom-made entrance reducer or with a piece of wood or sponge.
- Ensure that the hive is tilted slightly forward and leave for a week before carrying out a full inspection (see Section 10).

Placing a package of bees

1. Place the newly arrived package in a cool, dark room (18–20 °C).
2. Sprinkle or spray some 1:1 sugar syrup on the screen surface.
3. Open up the empty hive and install the bees in the late afternoon.
4. Reduce the hive entrance to prevent robbing.
5. Give a sharp bang on the cage's floor so that the bees fall on to it in the cage.
6. Remove the cage's wooden cover.
7. Remove the internal feeder.
8. Remove the queen cage and check that the queen is alive.
9. Push a nail or small stick through the queen cage candy to help the bees release her. (Remove any cork or plastic cover before doing this.)
10. Remove five frames from the empty hive.

11. Place the queen cage between two comb frames in the centre of the brood box with the entrance facing slightly upwards, ensuring that this entrance is free of obstruction.
12. Tip the bees from the package into the hive in the space made by removing the frames. (The bees must be able to get to the queen cage.)
13. Gently replace the five frames taking care not to crush the bees.
14. Close up and inspect in a week or two to ensure that the queen is laying and the colony is developing (see Sections 9 and 10).

Hiving a swarm

1. There are many ways to remove a swarm from where it has gathered, and you must use your imagination in this.
2. Place the swarm in any type of box and place this in a shady position near to where you captured it.
3. Wait until the evening when all the bees will be inside.
4. Seal up the box (ideally you will have a purpose-built box with a mesh screen but usually you won't. Don't worry about this).
5. Convey the box to the apiary and open up or prepare an empty hive.
6. Place in it one frame of comb or foundation.
7. Tip the bees in one knock into the hive.
8. Gently place foundation frames in the hive with one frame of comb if you have it.

9. Close up the hive ensuring that the entrance is reduced.

10. Check the swarm in a week to ensure that the queen is laying and the colony developing.

Note:
You can tip the swarm on to a ramp leading up to the hive entrance for effect! It impresses everyone and is a display that you can put on to convert doubtful neighbours.

SECTION 3:
DRIFTING

Drifting is a problem associated with most apiary sites and occurs when honeybees enter the wrong hive. This can cause some hives to be depleted of foragers, affecting their honey production capability, and it can spread disease. Loaded foragers not exhibiting 'robber' flight characteristics will usually be readily accepted into a strange hive. Drifting is a bigger problem than most beekeepers realise and many ascribe other problems, such as disease, to hives depleted by this phenomenon. The best way to control drifting is by prevention.

Factors causing drifting

- Hives placed in straight rows.
- Prevailing winds.
- Identical hives.
- Hives at the ends of rows are more conspicuous than hives within rows.
- Hives in the front and back rows are more conspicuous than middle-row hives. Bees will therefore more readily enter the outer rows of hives at the ends of the rows.
- New foragers are particularly prone to drifting.

Control after drifting has occurred

- Swap the position of the weak and strong hives.
- Transfer frames of bees and brood from strong hives to weak hives.

Part A

Prevention measures

- Arrange hives in irregular patterns, such as a U shape, with the entrances facing outwards.
- Arrange hives in pairs with 2–3 m between the pairs.
- Ensure that there is at least 3 m between rows of hives.
- Paint different-coloured symbols on the hive entrances.
- Place or grow landmarks in the apiary.
- Place artificial or natural wind breaks across the line of the prevailing wind.

Pointers

- Drifting can cause large variations in hive yields in an apiary. These variations are often put down to other reasons and the wrong corrective action taken (bad queen, etc.).
- Drifting is a bigger problem than most beekeepers realise.
- Very few beekeepers control or prevent the problem.

SECTION 4:

SITING ORGANIC APIARIES

Some beekeepers may wish to produce organic honey for various reasons, including the fact that there is a money premium involved. Organic rules are fairly strict and, although they may vary in detail between countries, the overall rules seem to be standard. Whether they are adhered to or not in some countries is another matter! The following basic standard rules are used by most countries.

When siting organic apiaries you must:

- Ensure enough natural nectar, honeydew and pollen sources for bees and access to a water supply.
- Be sure that, within a radius of 2 miles from the apiary site, nectar and pollen sources consist essentially of organically produced crops and/or spontaneous vegetation and crops not subject to the provisions of the regulations but treated with low environmental impact methods that cannot significantly affect the beekeeping production as being organic.
- Maintain enough distance from any non-agricultural production sources possibly leading to contamination (for example, urban centres, motorways, industrial areas, waste dumps, waste incinerators, etc.). The inspection authorities or bodies will establish measures to ensure this regulation.

Note:
The above requirements do not apply to areas where flowering is not taking place or to when hives are dormant.

The pedigree of the bees is not important, and so the following sources of bees are acceptable:

- Colonies in existing organic hives.
- Colonies confined to brood chambers covered only by a queen excluder.
- Divided colonies from conventional hives on brood combs only.
- Package bee colonies.
- Nucleus colonies (nucs).
- Captured wild or migratory swarms on brood comb only.

Note:
The treatment of diseases and various manipulations are also covered by organic regulations. The treatments available and methods used are constantly evolving, and so reference to these matters should be made at the time.

Part B

Crops, trees and plants for bees

Introduction

This field guide doesn't pretend to be a comprehensive guide to flowers, trees and crops but, if you are setting up apiaries either for the first time or for expansion, or if you have decided to take your bees to pollination contracts or to areas of flowering crops for the honey, then you need to know which flowers offer nectar and pollen, and when. (Remember that without early pollen sources your hives will not build up swiftly to take advantage of the honey flow.) This guide offers a quick but basic checklist.

Flowers can vary immensely in their offering of nectar and conditions need to be 'right' for a good harvest. Generally, once you have been in an area for a few years, you will get to know the ideal conditions for a good harvest, but when you first move into an area it is wise to tour round to see what there is. Don't start apiaries in areas where there is no forage.

This section offers a useful, handy guide to the flowering plants in your neighbourhood and indicates when they flower, also giving an indication of their nectar flow so that you can get started in your new area with some degree of confidence. For more comprehensive information on flowers, trees and crops please see the further reading list.

SECTION 5:
TREES FOR BEES

Trees are a very useful source of nectar and pollen for bees, especially in urban areas, such as city streets, parks and gardens.

Key

N–	**Weak nectar producer; nectar only produced when weather is good enough.**
N	**Nectar producer.**
N+	**Good nectar producer.**
N++	**Profuse nectar producer.**
P	**Pollen producer.**

Part B

Fruit trees

All fruit trees (see Table 1) are good sources of early pollen and nectar but some are short lived. Many very early fruit trees suffer from a rainy, windy spring when the petals are blown off.

TABLE 1: Fruit trees

Tree	Flowering Period	Type of Producer
Almond (*Prunus dulcis*)	Early spring	N++
Apple (*Malus pumila*)	April to late May	N
Cherry (*Prunus cerasus*)	May	N+
Medlar (*Mespilus germanica*)	May	N
Peach (*Prunus persica*)	April–May	N+
Pear (*Prunus communis*)	March–April	N–
Plum (*Prunus domestica*)	Early April	N+
Quince (*Cydonia oblonga*)	Spring	N+

TABLE 2: Other trees (including hedgerow trees)

Tree	Flowering Period	Type of Producer
Acacia (*Acacia dealbata, A. longifolia*)	Late winter	N
Alder (*Alnus glutinosa*)	January–March	P (early pollen)
Blackthorn (*Prunus spinosa*)	March–May	N P (early pollen)
Cherries (*Prunus* spp.)	March–April	N P
Eucalyptus spp.	August–September	N (in some areas of Europe 2 flowerings annually)
False Acacia (*Robinia* spp.)	May–June	N P
Hawthorns (*Crataegus* spp.)	May	N (can be erratic to very good) P
Hazel (*Corylus avellana*)	March–April	P
Hollies (*Ilex* spp.)	May–June	N P (useful in the June gap)
Horse chestnuts (*Aesculus hippocastanum*)	April–May	N+ P
Indian bean tree (*Catalpa* spp.)	July–August	N P
Judas tree (*Cercis siliqua*[illegible])	April–May	N P
Limes (*Tilia* spp.)	Many [illegible] flower spring–summer	N++ (if conditions right)
Maples (*Acer* spp.)	Spring	N P
Mountain ash (*Sorbus aucuparia*)	Spring	N P
Sweet (Spanish) chestnut (*Castanea sativa*)	July	N+ P
Sycamore (*Acer pseudoplatanus*)	May	N+ P
Tree of Heaven (*Ailanthus altissima*)	July–August	N (usually in towns)
Tulip tree (*Liriodendron tulipifera*)	June–July	N
Whitebeam (*Sorbus aria*)	May–June	N P

SECTION 6:
CROPS AND WILDFLOWERS FOR BEES

Crops

The list of crops pollinated by bees is huge, especially if you take into consideration the fact that at least one third of what you eat has been bee pollinated. Table 3 gives you an indication of some of the main crops that you are likely to encounter and which can give good pollination returns – whether from honey or from pollination contracts.

TABLE 3: Crops

Crop	Flowering Period	Type of Producer
Alfalfa (*Medicago sativa*)	July–August	Can irritate bees
Asparagus (*Asparagus officinalis*)	May–June	
Blackberries (*Rubus* spp.)	May–June	
Borage (*Borago officinalis*)	June–frost	Often a set-aside crop
Buckwheat (*Fagopyrum esculentum*)	July–August	Good for coughs
Chives (*Allium schoenoprasum*)	May–September	
Clovers (*Melilotus* spp. and *Trifolium* spp.)	May–August	
Cucumbers (*Cucumis* spp.)	May–June	
Melons (*Cucumis melo*)	June–frost	
Mustards (*Brassica* spp.)	Apr–May	
Oilseed rape (canola) (*Brassica* spp.)	May–June	
Pumpkin (*Cucurbita pepto*)	June–frost	
Raspberry (*Rubus*)	May–June	
Sainfoin (lucerne) (*Onobrychis viciifolia*)	May–July	
Soybean (*Glycine soja*)	July–October	
Sunflower (*Helianthus annuus*)	June–September	
Sweetcorn (maize) (*Zea mays*)	June–July	
Thymes (*Thymus* spp.)	June–July	
White sweet clover (*Melilotus alba*)	June–August	

Wildflowers (many can be used as garden plants)

There are literally thousands of wildflowers (weeds) that produce nectar and, in some European countries, 'wildflower' honey is a major spring harvest. Bees love weeds and weeds provide delicious honey. Many of them can be used for the garden. Unfortunately in the countryside where monocrop farming is the norm, weeds are shot on sight but, in areas of set-aside, great harvests can be obtained. Table 4 lists some of the commonest wildflowers that you will encounter and that provide nectar and, in many cases, pollen.

TABLE 4: Wildflowers

Flower	Flowering period
Ajuga (*Ajuga reptans*)	Mid-spring
Alyssum (*Lobularia maritime*)	June–September
Basil (*Ocimum basilicum*)	June–July
Blue thistles (*Carduus* spp.)	July–August
Borage (*Borago officinalis*)	June–frost
Catnip, cat mint (*Nepeta* spp.)	June–September
Chickweed (*Stellaria media*)	April–July
Cotoneasters (*Cotoneaster* spp.)	July–August
Dandelion (*Taraxacum officinale*)	April–May
Elder (*Sambucus Canadensis*)	June–July
Fireweed (*Epilobium angustifolium*)	July–August
Germander (*Teucrium chamaedrys*)	July–August
Goldenrods (*Solidago* spp.)	September–November
Heather (*Calluna vulgaris*)	July–August
Hemp (*Cannabis sativa*)	August
Henbit (*Lamium amplexicaule*)	April
Ivies (*Hedera* spp.)	September–November
Knapweed (*Centaurea nigra*)	July–August
Lemon balm (*Melissa officinalis*)	June–September
Leopard's bane (*Doronicum cordatum*)	April–May
Mallow (*Malva alcea*)	June–September
Marigold (*Calendula officinalis*)	June–September
Oregano (*Origanum vulgare*)	June–September
Peppermint (*Mentha piperita*)	June–September

Poppy (*Papaver somniverum*)	May–June
Snowdrop (*Galanthus nivalis*)	March–April
Star thistles (*Centaurea* spp.)	July–September
Viper's bugloss* (*Echium vulgare*)	June–August
Virginia creeper (*Parthenocissus quinquefolia*)	July–August
White mustard (*Sinapis alba*)	June
Winter aconite (*Eranthis hyemalis*)	March–April
Yellow crocus (*Crocus vernus*, syn. *C. aureus*)	April

*The flower of viper's bugloss (*Echium vulgare*) provides protection for the nectar inside from vaporisation (when it's hot) or flushing away (when it rains). Additionally, this plant produces nectar throughout the day, unlike most plants which produce nectar for a short period of time.

SECTION 7:

HONEYDEW SOURCES FOR BEES

Remember the trees and bushes listed in Table 5 as sources of honeydew. Honeydew honey is delicious and a useful source of income in Europe, if you can get it. It produces best when pests, such as leafhoppers and aphids, thrive – often in times of drought and pests are allowed to build up. In Germany, ant colonies are sometimes taken into the forests to 'guard' and encourage the aphids that produce the honeydew. Most of the trees listed below are minor sources but some, such as the beech trees of New Zealand and the silver fir of Germany, produce huge crops and are major exports. The cork oak trees of Spain produce excellent honeydew during periods of drought.

TABLE 5: Honeydew sources for bees

Source	Notes
Aleppo pine (*Pinus halepensis*)	
American elm (*Ulmus Americana*)	
Apple (*Malus sylvestris*)	
Apricot (*Prunus armeniaca*)	
Ashes (*Fraxinus* spp)	
Beeches (*Fagus* spp.)	
Black beech (*Nothofagus solandri*)	New Zealand honeydew source
Bulgarian fir (*Abies borisiiregis*)	
Cherries (*Prunus* spp.)	
Gooseberry (*Ribes uva-crispa*)	
Greek fir (*Abies cephalonica*)	
Hawthorns (*Crataegus* spp.)	
Hazels (*Corylus* spp.)	
Hickories (*Carya* spp.)	
Larch (*Larix deciduas*)	
Limes (*Tilia* spp.)	
Maples (*Acer* spp.)	
Norway spruce (*Picea abies*)	
Oaks (*Quercus* spp.)	Cork oaks source of Spanish honeydew
Peach (*Prunus persica*)	
Pear (*Pyrus communis*)	
Pines (*Pinus* spp.)	
Plum (*Prunus domestica*)	
Red beech (*Nothofagus fusc*)	New Zealand honeydew source
Rowan (*Sorbus aucuparia*)	
Scots pine (*Pinus sylvestris*)	
Silver fir (*Abies alba*)	Source of German forest honey
Turkish pine (*Pinus brutia*)	
Weeping willows (*Salix* spp.)	
Willow (*Salix alba*)	

SECTION 8:
BEE PLANTS AND SHRUBS FOR THE GARDEN

(see Table 6)

Key	
N–	**Weak nectar producer; nectar only produced when weather is good enough.**
N	**Nectar producer.**
N+	**Good nectar producer.**
N++	**Profuse nectar producer.**
P	**Pollen producer.**

TABLE 6: Shrubs

Shrub	**Flowering Period**	**Type of Producer**
Berberis spp.	April–July	N+ P
Brooms (*Cytisus* spp.)	February–April	N P
Buddleia spp.	May [illegible]	N P
Calico bushes (*Kalmia* spp.)	June	N P
Ceanothus spp.	May–September	N++ P
Chaenomeles spp.	February–April	N P
Cherry laurel (*Prunus laurocerasus*)	April	N P
Clematis spp.	July–December	N P
Cotoneaster spp.	June–July	N+ P
Daisy bushes (*Olearia* spp.)	July–August	N P
Deutzia spp.	July–August	P
Firethorns (*Pyracantha* spp.)	May–June	N P
Flowering currants (*Ribes* spp.)	April	N P
Gorses (*Genista* spp.)	January–April	NP
Hebe spp.	June–September	N+ P
Honeysuckles (*Lonicera* spp.)	Late winter onwards	N P
Ivy (*Hedera helix*)	September–November	N P

Mahonia spp.	Winter–spring	P (good early pollen)
Mock oranges (*Philadelphus* spp.)	June–July	N P
Myrtus communis	Late summer	N– P
Parthenocissus quinquefolia	August	N+ P
Perovskia atriplicifolia	August–September	N++ P
Portugal laurel (*Prunus lusitanica*)	June	N P
Potentilla fruticosa	July–August	N P
Rock roses (*Cistus* spp.)	May–July	N P
Rosa spp. (wild roses and *R. rugosa*)	March–April	P
Rosemary (*Rosmarinus officinalis*)	April–May	N+ P

Nectar-producing garden flowers

Shade plants

Cotoneasters, cranesbills, foxgloves, lungwort, sweet box.

Partial or full sun

Bergamots, buddleias, catmint, columbines, cornflowers (perennial), cranesbills, echinacea, escallonia, evening primrose, globe thistles, heathers (heath), heathers (ling), hebe, Japanese roses, laburnum, lavenders, michaelmas daisies, mulleins, rosemary, roses, rudbeckia, sage, scabious, sea holly, stonecrop, thymes, valerian, verbena, wallflowers (perennial), wisteria, yarrow.

Cool periods

Hellebores, mahonia, sweet box, witch hazel (bees will only fly in winter if the temperature is high enough).

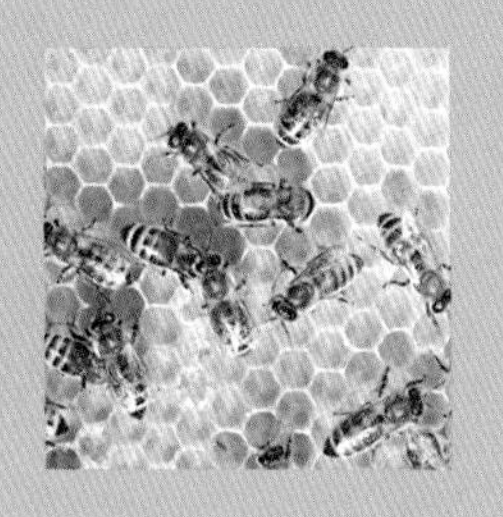

Part C

Troubleshooting guide to field operations

Introduction

This part provides both a series of troubleshooting guides for problems commonly associated with colonies and a series of checklists for manipulations designed to overcome these and other problems.

Without doubt, there are many more methods of achieving the aim, both 'secret' and well known, but I have carried out each of the methods listed here and, while nothing is certain with bees, these methods carry with them a high degree of success. You can use the information given here with confidence. It has all been tried and tested.

In many situations, most beekeepers will be able to recognise when things are going well – when the hive looks 'normal' with the bees going about their business in a manner appropriate to the time of year and the weather. If you can recognise these signs in your apiary, even from a distance, you will immediately know if something is wrong: bees fighting; bees flying busily from all the hives except one; piles of dead bees in front of the hive entrance; when you open the hive you see many eggs in single cells – and so on.

It is when you see something amiss that you should consult this section of the Field Guide for confirmation of the problem and suggested solutions. Most situations are covered but those problems caused by disease are referred to in Part H.

SECTION 9:

HIVE-ENTRANCE INSPECTION/ TROUBLESHOOTING GUIDE

Much can be ascertained about the health of a colony by observing the hive's entrance. This type of inspection may be useful if you do not want to disturb the colony for some reason – e.g. you have just set up a nucleus or you have just introduced a queen. If you notice anything listed in Table 7, check further.

TABLE 7: Hive-entrance inspection/troubleshooting

Observation	Interpretation
Bees fighting at the entrance	Robbing (see Section 13)
Pile of dead bees at the entrance with equal decomposition of bees in pile	Poison (see Section 17); can also indicate starvation
Pile of dead bees at entrance. Bees at bottom of pile decomposed, bees at top still moving	Virus disease (see Section 48)
Dead drones at the entrance/ drones being removed	Period of dearth present or coming
Bees unable to fly or staggering/ moribund on hive or entrance board	Virus disease (see Section 48); can also indicate starvation
Mummified larvae littering entrance	Chalk brood (see Section 43)
Faeces spotted around the entrance	Dysentery (see Section 47)
Strong, foul smell coming from hive	American foul brood (AFB) or European foul brood (EFB) (see Sections 41 and 42)
Dead larvae being thrown out but not carried away (larvae may have been sucked dry)	Possible starvation
Dead larvae being carried out	Brood disease (see Part H)
Pollen being carried in	Usually indicates a healthy colony
Many bees flying at entrance. No fighting. Bees tend to be facing the hive and appear to be bobbing up and down	Young bee play/orientation flights

Many bees issuing from hive in a swirling ascending mass	Swarm emerging (see Section 18)
Many drones flying	Normal for a colony between around 2pm and 4pm
A regular column of ants entering and leaving the hive. No bees entering, leaving or on guard	Hive or nuc empty of bees. Put your ear to the side of the hive and give it a sharp knock. If silent, the hive/nuc is empty. Ants often move in after the hive has been robbed out

Note:
These interpretations are a guide only and cannot take the place of a full internal hive inspection. If all the hives are busy but one isn't, check it out.

SECTION 10:

HIVE-INSPECTION TROUBLESHOOTING GUIDE

Before opening a hive, decide exactly what you are going to look for. Every general hive inspection is essentially a health check and should include a look at the following:

- Queen and/or eggs and very young worker brood. One egg per cell (see Section 11 for queen/brood problems).
- A good brood pattern (see 'The brood pattern: pointers' below and refer to Section 11).
- Sufficient stores of honey and pollen, especially for the winter/dearth months which can include late spring/summer in many areas.
- Sufficient room in the brood nest and supers for the queen to lay and the workers to store honey and pollen.
- Presence of swarm cell(s)/supersedure cell(s) (see Section 18 for swarm control methods).
- Signs of disease (see Part H for a comprehensive field diagnosis guide).
- Cleanliness of the hive floor. Any large build-up of debris should be cleaned away or a new floor put on. This can indicate a problem with the colony so look for any other signs of trouble (see Part H). Ensure that there is no build-up of water in the corners, etc. – preferably use a stainless-steel mesh floor to avoid this.

The brood pattern: pointers

- Inspection of the brood pattern is one of the simplest and most accurate ways of assessing the health of the colony.
- The presence of sealed brood indicates what was happening 9–21 days ago.
- Inspection of young brood and eggs provides a more up-to-date assessment of the health of the colony.
- Bear in mind that not all problems with the brood nest can be attributed to a faulty queen. Pesticide poisoning (see Section 17) can result in several brood-nest problems, the symptoms of which are often similar to those of disease (see Section 11 if you find a problem with the brood nest or you think there is a problem with the queen).

Note:
Wax moth larvae may be present in a hive, but not in any numbers. If, however, there is any sign of actual wax moth damage, another problem may exist and further investigation is necessary (see Section 40).

SECTION 11:

QUEEN/BROOD-NEST TROUBLESHOOTING GUIDE

(see Table 8)

The queen bee is your main production unit. The health and functioning of the colony depend entirely upon her health. A good, gentle, productive queen can transform your beekeeping. The whole tenor of the hive, from temper to honey production, is affected and you, as the beekeeper, can control this by regular re-queening with known, gentle and productive queens. A bad queen should not be tolerated. Whenever you inspect your colony, *always* ensure that you have a laying queen present.

TABLE 8: Queen/brood-nest troubleshooting

Problem	Cause	Treatment
No brood present	No queen/failed queen	Re-queen or unite colony (ensure it is not a time of a lull in egg laying – e.g. a cold winter period)
Sealed brood only; no eggs	Colony swarmed	Check in three weeks for eggs or young brood
Drone brood only; one egg per cell	Drone-laying queen (queen failure)	Re-queen/unite colony with a strong colony
Drone brood only, often in worker cells. Eggs not at base of cell	Laying workers	See Section 12
Mix of drone brood in worker cells, normal capped brood, several eggs in some worker cells	Laying workers	See Section 12
No brood. Small queen, excitable on comb	Virgin queen, delayed mating or not yet mated	Check for eggs in one week
	Newly arrived postal queen	Check for eggs in one week

Part C

Supersedure cell(s) formed after queen introduction; common	Cause unknown	Remove cell(s). Can be cut out and put in a queenless nuc
	Badly mated queen	Check brood pattern if bad, allow supersedure
	Cells were present before introduction	Destroy cells
Queen in introduced cell dies	Not fed by workers or cage balled	Laying workers may be present (see Section 12)
Introduced queen killed after release	Old queen present	Remove old queen prior to introduction
	Unnoticed virgin present	Leave her to mate or kill her and re-queen (see Section 28)
	Laying workers present	See Section 12
Small but good brood pattern	Newly mated queen	Inspect again in two to three weeks to see if she is maintaining her egg production
	Slow laying queen	Re-queen or accept situation (see Section 28)
	Not enough bees to look after brood	Allow colony to build up or, if serious, add more bees or unite (see Section 15)
	Not enough room for queen to lay	Provide comb/ clear brood nest
Poor brood pattern (larvae of different ages together) by the queen	Inbreeding, leading to removal of diploid drones and re-laying	Re-queen if serious. It is pointless to maintain an inbred queen except as part of a breeding research programme
Swarm cells present	Colony preparing to swarm	See Section 18
Two queens present	Supersedure queen and daughter	Leave alone if no fighting. Old queen will disappear. Or split hive
	Swarm(s) waiting to go	Virgin(s) will probably leave with swarm (see Section 18)

Notes:

Many queen problems are caused in the following ways:

- *Queens damaged or killed during manipulations, especially rolling/crushing a queen while removing frames in a tight hive and also by queens flying off a removed comb. It does happen!*
- *Queens introduced while the old queen is still present.*
- *A queen introduced when laying workers are present.*
- *Inability of beekeepers to find queens and so making the wrong assumptions.*

SECTION 12:

LAYING WORKERS

(see photographs 3, 4 and 5)

Indications

- You can see only drone brood, particularly in worker cells. This is very noticeable.
- Eggs adhering to the sides of the cell. Worker bees do not have the long abdomen of the queen and so cannot reach the bottom of the cell to lay the eggs.
- Several eggs in the same cell. Laying workers compete to lay eggs and will lay in cells that already contain eggs. (In certain circumstances – e.g. in the spring – a normal queen may lay more than one egg in some cells, but usually at the bottom of the cell.)
- Eggs positioned some way up in the cell wall.
- A scattered laying pattern. Normal queens work from the centre of the brood nest outwards. Laying workers will operate in a more random manner.
- Eggs laid on and around larvae. Again, laying workers will compete with each other and lay in cells already occupied.
- Larvae neglected by nurse bees. The laying workers are eggs that will turn into diploid drones that are of no use to a colony.
- No queen.

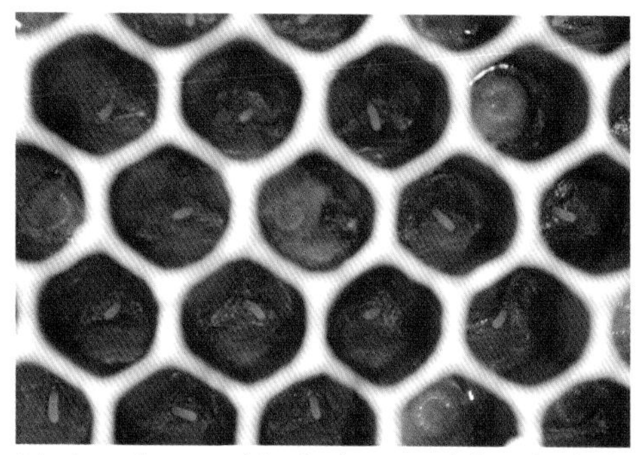

3. *Laying workers: normal situation (one egg at the base of each cell)*

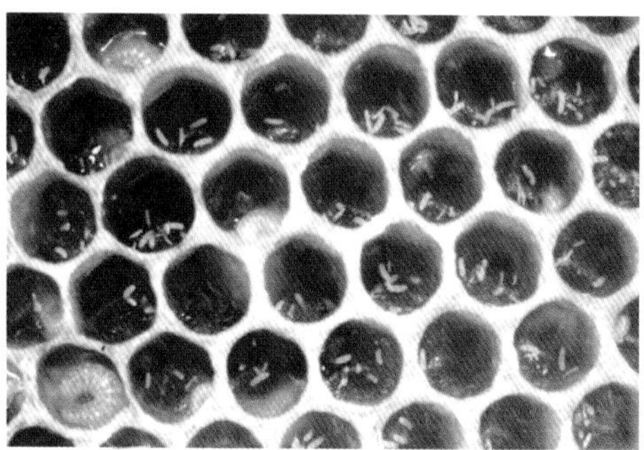

4. *Laying workers: multiple eggs in worker cells (evidence of laying workers and no queen)*

5. *Laying workers: the result – drone pupae in enlarged worker cells showing the domed cappings*

Treatments

There are two possible treatments that are usually successful. Before trying to correct the problem, decide whether it would be more worthwhile to disband the colony. If the problem is severe and the colony is very small, you may see wax moth damage. If it has gone this far it is probably best to disband the colony. I have found that, in most cases, it is not worth keeping the colony because it has 'gone' too far.

Treatment 1

1. Move the entire colony 200 m away (leaving a hive body with comb for the returning foragers on the original site).
2. Clear all the frames of bees from the hive.
3. Tap out all the eggs and larvae. Uncap all the drone brood and remove it.
4. Return the bee-less hive to its original site. (Most of the bees from the hive will return to the original site, but the laying workers lose their sense of orientation or will never have flown and therefore are unable to find their way back.[1])
5. Give a frame of eggs and brood, or queen cell, or introduce a queen in an introduction cage (see Section 28), or unite with a queen-right colony/nucleus on the same site as the original hive (see Section 15).
6. Feed if necessary using a frame feeder.
7. Close up and leave for an appropriate period – at least a week to see if the new queen has been accepted and is laying.

Treatment 2 *(best used if the colony is still large)*

1. Remove all the combs in the brood chamber (you are removing the laying workers).

2. Confine them without food for two days in a dummy hive or nuc.

3. Replace the combs after two days.

4. Introduce a laying queen (see Section 28).

5. Feed.

During the period of confinement, the laying workers' ovaries should return to their normal state and they become less of a threat to a newly introduced queen.

Notes:

- *Never try to re-queen a colony containing laying workers. The queen will probably be killed.*

- *Giving a colony of laying workers a frame of brood or a queen cell will not usually work.*

- *Never unite a colony containing laying workers to a queen-right colony. Your queen will probably be killed.*

Part C

SECTION 13:

ROBBING – PREVENTION AND CONTROL

Diagnosis

- Much increased activity at the hive entrance. Bees will be flying all around the hive and the hive may be covered with bees.
- Bees fighting at the entrance to a colony.
- Much debris, particularly wax particles, at the hive entrance caused by robber bees tearing down the comb to rob the honey.
- Many bees entering and leaving every small crack in the hive that is being robbed.
- A general air of confusion in the apiary.

Probable causes

- Exposure of nectar, honey or sugar in the apiary near to a weak hive, especially during a period of dearth. This can alert bees to robbing opportunities, but this does not always occur – it is usually the next cause that starts it off.
- A diseased and/or numerically weak colony (e.g. a nucleus) with stores that is unable to mount guard bees and that is discovered by scout bees looking for honey, especially in a period of dearth.

Control procedure

Once robbing starts it is very difficult to stop that day. It is often better to go home and hope for the best, as long as your neighbours are not being affected. However, the following actions may help to reduce it and may help the colony that is being robbed:

1. Block up all the cracks in the hive being robbed with grass, mud or paper or whatever comes to hand to prevent access by the robber bees.
2. Reduce the entrance of the hive being robbed to one bee space to allow even a weakened colony to defend itself.
3. If available, lean a glass screen or board across the hive entrance, allowing room around the sides for exit and entry. This confuses the robbers. *Or* place handfuls of grass or straw across the entrance, *or* swap the positions of the hive being robbed with the robber hive. This is a more drastic solution and you may decide to move the robbed hive out of the apiary completely (more than 1 mile away).

At the end of the day, robbing will cease. If the situation was serious, it may be worth moving either the robbed colony or the robber colony to a new location at least 1 mile away.[2]

Prevention

- Do not spill honey, nectar or sugar during periods of dearth.
- Ensure that colonies are kept disease free so that they do not become weak and an attraction to robbers.

- Ensure that small colonies (e.g. nuclei) have small entrances, especially during dearth periods. One bee space is best. (Ensure adequate ventilation in hot weather.)
- Ensure that all hives and nuclei have crack-free boxes.
- Avoid manipulating colonies during dearth periods.
- Weak colonies that are slow to build up can be united with strong colonies that are not required (see Section 15). Check for disease first (see Part H).

Notes:

- *Robbing disrupts and can destroy colonies, increases the aggressiveness of the bees, and is highly instrumental in spreading disease, the most virulent of which is American foul brood (AFB).*
- *Small nuclei are especially vulnerable to robbing, so look after them well (especially the entrances), but always ensuring adequate ventilation.*
- *Italian bees are usually the first in.*

SECTION 14:
AGGRESSIVE COLONIES

(see Table 9)

Bees will defend their nest from predators, including beekeepers, but the degree of aggressiveness in this defensiveness can vary, depending on a number of factors:

- The race of the bees. Italians, Cecropians and Carniolans are the most popular races for comparative gentleness. Africanised bees and the Iberian bee are not noted for gentleness!
- The local environment (see below).
- Handling by the beekeeper. This is a huge problem, with beekeepers banging hives and frames around.
- Insecticides – another big problem in many areas (see Section 17).

A variety of factors may influence how aggressive a colony becomes. As a general rule, established colonies are more aggressive than small nuclei, and so any test of aggression should be made when a nuc grows into a colony. The degree of aggression also depends upon the beekeeper's perception. One person's aggressive bee…

TABLE 9: Causes of aggression and remedies

Causes of aggression	Remedy
Queen genetics (Moritz et al., 1987)	Re-queen with a queen from a gentle race/stock (see Section 28)
Hive being robbed by bees or wasps	Find and destroy the wasps' nest (see Section 13)

Part C

Hive being disturbed by large animals or vandalism. Cows scratching themselves on hives is a prime example	Re-site hive or fence off hives
Hive badly sited under power line/ near busy road/under dripping trees	Re-site hive[3]
Bad weather affects bees in a variety of ways	Work in good weather[4]
No honey flow	Check for sufficient stores
Beekeeper works a bad-tempered colony which then disturbs a nearby gentle colony	Always work the gentle colony first so as not to alarm the other colonies
Colony is queenless	Re-queen (see Section 28)
Bees affected by spray poisoning	See Section 17[5]
Large-scale use of insecticides within foraging range	Apply protective measures (see Section 17)
Small-scale use of garden insecticide upwind of the apiary	Probably no action required[6]

Notes:

- *There is no direct correlation between aggressiveness and good nectar collection.*
- *Rough handling and the crushing of bees by the beekeeper will also cause a high degree of defensiveness, even in normally quiet colonies.*
- *In some areas of the USA, the most common cause of aggressiveness in colonies is the persistent attention and predation of skunks and their families.*

(See also Bosch and Rothenbuhler, 1974; Breed et al., 1990.)

Inspecting a colony of aggressive bees

No beekeeper, however experienced, likes to be pasted by their bees, and it is never pleasant to end the day worse the wear from stings because of a colony of very aggressive bees. Some colonies showing extreme aggressiveness are a daunting prospect at times. So make it easy on yourself. Carry out the following procedure if you have a colony that is consistently aggressive:

1. In the early evening of a good flying day, seal up the entrance to the hive in question. Do this quickly using a sponge strip or grass, etc. Have this ready to use immediately.
2. Move the entire hive 10–15 m to one side.
3. Place a hive body, lid and floor on the original site. In this, place one or more combs of honey and pollen (depending on the size of the colony) to collect the returning foragers (the stinging bees).
4. Leave for an hour or two or, better still, overnight for things to settle down.
5. The next day, manipulate the aggressive colony. Most of the stinging foragers will be in the dummy hive.
6. Kill the aggressive hive's queen, which can then be re-queened (see Section 28) or united (see Section 15) with a gentle colony.
7. The foragers in the dummy hive can be given a frame of eggs and young brood or a queen cell (from a gentle colony) so they can raise a queen of their own. (This usually works, but see also Section 26 for field queen-rearing methods.)

Notes:

- *If at any time you bungle the operation and the bees get out of hand, cover everything up, walk away and start again another day.*
- *With some colonies smoke can often make things worse. If so, try not to use it, or make minimal use of it.*

SECTION 15:
UNITING COLONIES AND NUCLEI

This is a simple procedure in most cases, and there are many ways of doing this, but precautions do have to be taken. Simply mixing all the frames into one box will often work if there is a good flow on but, by not taking care, many foragers could be lost to no purpose and you might lose a good queen. The aim is to unite two colonies or nucs with the minimum of fighting and with no queen loss. The two methods outlined below usually work well.

Procedure 1

1. If both colonies are queen-right, find and kill the queen you don't want.

2. Remove the lid of the queenless colony. Place a sheet of newspaper over the hive and make a few slits in it.

3. Place the queen-right or stronger colony over this, ensuring that the upper unit has no entrance/exit.

4. Leave for 24 hours. Then adjust the boxes as you require them. The bees will have mingled.

Procedure 2

1. If both colonies are queen-right, find and kill the queen you don't want.

2. Open up the queenless or smaller hive.

3. Lift the hive to be placed on top and spray the underneath with a non-toxic room-freshener spray. (Don't use too much – just a very quick spray.)

4. Spray the top bars of the hive on the ground.

5. Place the top hive on top and leave alone for 24 hours. The bees will have mingled.

Notes:

- *Some beekeepers object to spraying any chemicals into the hive. However, this is a very fast and effective method of uniting colonies (useful for commercial beekeepers).*
- *You could use a sugar syrup spray instead but you need to be aware of the possibilities of attracting robbing.*

Procedure *(when uniting nuclei)*

1. Find and kill the queen you don't want (if there is one).
2. Spray all the bees in both nucs with a sugar syrup of a 1:1 mix (i.e. 1 k of sugar to 1 l of water).
3. Place them together in the same brood chamber (place the frames of brood together).
4. Feed if required using a frame feeder, especially during periods of dearth.
5. Close up and leave for 24 hours.
6. After 24 hours, check to ensure that the queen is still alive and well and that there is no fighting. Bees should be flying normally from the nuc.

Notes:

- *Remember that nuclei are prone to robbing (see Section 13) and to overheating in hot weather. If overheating occurs they may abscond, even abandoning brood, so ensure that there is adequate ventilation in the nuc in the form of openings covered with gauze.*

- *If the queenless nuc is much stronger than the queen-right nuc, then prior to step 3 above, cage the queen in a standard introduction cage with a candy-release block. This will give the bees time to settle down and accept the queen without balling her.*
- *It is always best to combine colonies during periods of abundant forage. Less fighting ensues as the forager bees are too busy collecting nectar, etc. If fighting does occur, spraying with a hose, if available, helps calm them down.*[2]
- *Some beekeepers advocate the use of flour when uniting nuclei instead of sugar syrup.*
- *Don't unite a colony of laying workers with another colony (see Section 12).*

6. *Staples are easy and fast and they don't upset the bees as much as some reports say. You get holes in the woodwork, though, which can allow the entry of moisture*

SECTION 16:
MOVING BEES

This is an easy process if care is taken. Photographs 6, 7 and 8 show various methods of securing hives before a move that are easy to use. Straps are probably the most employed method of securing hives but can be the most expensive, especially if using metal-type ratchet straps. This section is for beekeepers who need to make small-scale moves, perhaps in a car or small van, and who do not have specialist equipment or expensive straps. For small moves, bits of wood and nails and/or lengths of rope can all be used by the field beekeeper if nothing else is to hand.

7. *Wooden laths are cheap and easy but these do upset the bees more than staples when attaching them. They are ideal for the 'bush beekeeper' and can be made up and applied in the field in the absence of other fastenings*

Procedure for moving bees within the apiary

(don't unless you have to)

8. *Spring clips are secure but they can get in the way when not in use and so become irritating*

1. Few precautions are necessary but, to make a success of it, you must pay attention to what is going on.

2. Plug the entrance with grass or a sponge to make it difficult for the bees to escape for a while.

3. Move the hive and leave the block in place.

4. Place a brood box with some comb in it and a floor and lid on the old site to collect foragers from the field (if any) and those returning from the new location once they do manage to open up an exit in the grass.

5. When flying has ceased, place this 'dummy' hive on the hive in the new location. Repeat the next day. Or move the hive a metre/yard each evening until you reach your destination.

Procedure for moving colonies short distances

(up to 2–3 hours)

1. If cool or at night, plug the entrance and any cracks, secure the hive and go. If you have hive screens, these are ideal and there is no need to add an empty super.

2. If the weather is very hot and you are moving during the day, take the same precautions as for moving bees over long distances and for retaining bees in the hive (see Section 17).

3. Use common sense with regard to distance and temperature. Bees are hardy creatures but can suffer stress during moves.

Moving colonies long distances

1. If you move bees by day during hot weather and/or it is going to be a long journey, ensure that the bees have plenty of ventilation, room and water. Follow the advice given for retaining bees in the hive in Section 17.

2. If you load up and move at night only, then you have no need to close up the hives or to cover the hives with a net. Just load and go.

Pointers

- It is best to seal up hives after flying has ceased for the day to include all the foragers. This is not essential, however. If you have to move quickly by day because of spraying, etc., just do it. You will lose foragers but that is not catastrophic. This is better than losing the whole colony to spray poisoning.

- Permanently fixed clips on hives often get in the way when you are not using them, and fixed-floor hives make many manipulations (such as reversing hive bodies) impossible. Many beekeepers try these devices but soon remove them.

- If moving bees in an enclosed vehicle and some bees escape, they will rarely, if ever, sting. They want out, so don't worry too much about bees flying around in the car.

SECTION 17:
SPRAY-DAMAGE PROTECTION

As a beekeeper you should engage with the rest of the agricultural community in your area. (In urban areas, ensure that the local council or parks authority knows about you.) Other farmers use insecticides and herbicides (both of which affect bee colonies) as a normal part of their farming activities, and they are not going to stop this perfectly legal activity just because they have a beekeeper in their midst. They will, however, accommodate and liaise with the beekeeper who engages with them and joins a local spray-protection scheme. These schemes ensure that the local farmer will alert beekeepers to any spraying well in advance so that the beekeeper can take the necessary precautions to protect their bees. Most beekeeping associations operate such a scheme, but it is also worth letting your neighbouring farmers know who you are and that you keep bees.

9. *Spray damage: thousands of dead bees at the entrance*

Assuming that you know when insecticides and other substances toxic to bees will be sprayed near to your hives, there are various methods of limiting the number of bee deaths and thus saving your colonies. Therefore, if there is a spray-warning scheme in your area, register with it. Once the bees have been poisoned (see Photograph 9) there is little you can do except to try to ensure it doesn't happen again.

10. *Hive made up for protection, entrance blocked with gauze. Empty super on top for room. Gauze top for ventilation. Loosely placed roof*

Procedures

1. Move the colonies out of the area (the best method), but do not move the bees to another area that is about to be sprayed. If this is not in your area, you may not have been notified by the authorities or the local farmers.

2. Move them back when the period of residual toxicity is over. You can ascertain this from the farmer or the local agricultural extension office. If this is impossible, the bees must be confined or their flights limited during the period.

3. Block the entrance with a bee-proof net or gauze (air must be able to enter).

4. Provide plenty of space in the hive by placing an empty half-super on it.

5. Provide adequate ventilation. Place a bee-proof net or gauze over the empty super. Put a board/lid over this, very loosely and propped up above the gauze at each corner (see Photograph 10).

Note: if the period of confinement is to be more than one day, or it is very hot, or both, then:

6. All of the above, plus a frame feeder with sugar syrup and a water container or wet sponge in the top empty super. (If they have sufficient stores, omit the feeder.)
7. Place wet sacking over the entire hive. Re-wet this periodically (see Photograph 11).
8. If possible, provide shade for the hives, using shade boards, branches or anything that comes to hand.
9. Keep the period of confinement as short as possible.

11. *Hive in a 'sack tent' – cheap, effective protection*

Note:

- *The important thing to remember is that, if you confine bees, you must provide them with adequate* room, ventilation *and* water *(and* food *if required).*[7]
- *Research shows that simply covering the hives with wet sacking, tent fashion (and keeping it wet), without blocking the entrance, significantly reduces bee mortality (Crane and Walker, 1983).*

If spray damage has already been caused

Spray damage can occur when either you as a beekeeper haven't informed anyone you are keeping bees in the area (see above) or if a farmer fails to notify the local spray scheme about spraying activities – or if they use illegal sprays – and this happens!

If damage has already been caused, an assessment of your colonies must be made, as follows.

Colony loses its foragers – pollen not carried into the hive:

1. Allow the colony to build up. Feed if necessary.
2. Apply protective measures if toxins are residual or more spraying is to be carried out *or* move the colony(s).
3. Add more bees or unite the colonies (see Section 15 and the pointers and notes below).

If poisoned pollen has been stored and brood and nurse bees are affected (brood continuing to die long after the event, colony not recovering, etc.):

1. Check the health of the queen.
2. Move the colonies to a safe area.
3. Remove combs of pollen.
4. Feed if required.
5. Watch colony development carefully. If all is well, add more bees or unite (see Section 15).

See also the pointers and notes below.

Pointers

- If poisoned pollen has been stored in the hive, then this must be removed. If the frames of pollen have brood in them, move the frames to a separate hive until the brood has emerged. Then shake these bees into the original colony and wash out the frames of pollen (see the notes below). Feed the original colony with pollen or substitute to reduce the number of pollen sorties from the hive.
- Pesticide poisoning may cause colonies to become stressed and prone to diseases, such as sac brood (see Section 44), chalk brood (see Section 43) and European foul brood (EFB) (see Section 42). Symptoms may occur up to 6 weeks after the event.
- The loss of bees after pesticide poisoning may result in insufficient bees to look after the brood nest, thus leading to chilled brood (see Section 45) and a subsequent spotty appearance of the brood nest as the dead larvae are removed. Don't confuse this with AFB/EFB (see Sections 41 and 42 and parasitic mite syndrome (PMS), Section 52).

Notes:

- *Combs containing poisoned pollen can be soaked in water for 2 hours, then the pollen washed out.*
- *It is not advisable to unite a poisoned colony with a healthy colony, for obvious reasons.*

Author's notes

1. *I have used treatment 1 several times and it failed only once. I think that the colony was too far gone and I should really have either united it to a strong colony (after getting rid of the laying workers) or disbanded it. The second method, advocated by Orosi-Pal in the 1920s, worked as well, but I tried it only once. It is easier, but takes longer overall.*
2. *One of my colonies (when I had only two) started robbing the other and the situation got totally out of hand. There were fighting bees all over the place, scaring the neighbours and worrying me. I took a hose to them. The fighting stopped. The bees settled on the underside of leaves on trees. As soon as I stopped the hose, they all came out and started again. And so it went on until dark. The neighbours were not impressed.*
3. *I moved a very aggressive hive from a location near to a busy road to a quiet position. The bees became less aggressive within a day.*
4. *I knocked against one of my quieter colonies in the apiary during a rainstorm and the bees went berserk. They came out after blood (and got it), ignoring the rain, and remained active and aggressive for over an hour. Mind you, they were Iberian bees!*

5. *During an inspection of colonies in Spain, I found that all the bees went for me immediately I opened the hives, even three colonies of Cecropian bees (Greek), which are normally not too bad. I used to expect a bit of trouble from my Spanish bees but this was something else. Smoke made them even worse. This behaviour lasted for two days. I connected it with some aerial spraying of olive trees in the vicinity during the two days before the inspection. There were few deaths due to the spraying. (I had been advised by the local extension vet that no precautions were necessary to prevent colony deaths.)*

6. *My colonies became very aggressive during a day when they were flying through an area in which a gardener was using a hand-spray insecticide on some garden plants. The gardener was upwind of the apiary. No bees (to my knowledge) were harmed, but I was.*

7. *Due to spraying, I had occasion to confine 10 colonies of bees for three days, in temperatures of over 30 °C. I gave them plenty of room and ventilation, together with ample water from spray bottles, and kept their water sponges full by tipping beer cans of water through the screen. I had minimal losses, just a lot of bad-tempered bees when I finally let them out.*

Part D
Swarm prevention and control

Introduction

Swarming is probably the most important bee-management problem in the production of honey or the providing of pollination services for beekeepers. Swarming is the natural method of increase of the honeybee colony – a sort of reproduction by increase rather than sexual reproduction. It is also somewhat of a mystery because the exact causes of swarming are not known. However, there are certain factors that are connected with swarm biology which serve as the basis of preventive manipulations. The control of swarming, once the honeybee colony has begun swarm preparations, involves extensive manipulation and critical timing on the part of the beekeeper before the situation takes control and swarms are lost.

Preventing swarming requires detailed knowledge of bee biology and an almost intuitive knack of when to carry out various manipulations in harmony with the flow of the bees and in good time to prevent what is a natural imperative on their part. Basically it requires time and knowledge on the beekeeper's behalf.

The prevention and, when necessary, the control of swarming is good bee management because, if your colony swarms, you will probably not obtain a surplus of honey from that colony, and neither is the colony fit to carry out pollination contracts where hives brimming with bees are required. Any management plan in any agricultural enterprise will tell you that an increase in your livestock numbers should be planned and not the result of bad management. If you understand swarming and make an effort to control it, your beekeeping will be more pleasurable and certainly more profitable. Let other beekeepers be slack in their swarm control and you will be able to pick up their bees for free (see Section 25)!

SECTION 18:

SWARM CONTROL METHODS

Look for early signs of swarming. The following are indications that a hive is heading towards a swarming situation:

- An overcrowded brood chamber with no room left for the queen to lay eggs.
- Lack of space in supers for honey storage.
- The laying of eggs in queen cups.
- The building of queen cells.

Other behavioural changes occur in colonies that are less easily observed, such as the following:

- Weight loss of the queen in preparation for flight (up to one third or one half of body weight).
- The field bees do less work and may congregate at the hive entrance and/or on the lower frames.
- More drones are reared.
- The queen lays fewer eggs.
- Prior to leaving the hive, the workers engorge themselves on honey and almost cease normal flight activity.

If you want hives brimming with bees to give you a good harvest, then you must try to prevent swarming. The following control manipulations are mainly aimed at reducing congestion within the hive which, if carried out in the spring, may help to reduce swarming, especially as part of an integrated swarm-control scheme. The last part of this section gives advice on how to recoup some of your losses from swarming (if this happens) by attracting other people's swarms (or even your own) to bait hives.

Methods

- Equalising colonies (see Section 19).
- Reversing hive bodies (see Section 20).
- Dividing colonies and making up nuclei (see Section 21).
- The artificial swarm (see Section 22).
- The Demaree method (see Section 23).
- Caging or removal of the queen for a period (often combined with queen replacement) (see Section 28).
- Using bait hives and swarm collection (see Section 29).

Pointers

- Some colonies will swarm whatever the beekeeper does. One noted beekeeper suggested that, with some colonies, the only way to prevent swarming is to blow them up!
- Some races of bees are much less prone to swarming than others.[1]
- Once you see capped queen cells in a large colony it is more than likely that your bees have already swarmed or are on the point of doing so, so look around for a virgin. If they haven't swarmed yet, then only desperate measures will prevent them.

Notes:

- *Regular re-queening with young queens plays an important part in reducing the incidence of swarming. Check out the figures – they are amazing!*
- *The two-queen management system (see Section 32) also reduces the incidence of swarming.*
- *Clipping queens' wings and cutting out queen cells are best carried out as part of an integrated swarm-control programme. On their own, even if successful, they fail to address the cause of the problem, work against the bees' instincts, can make the bees aggressive, reduce honey yields and frustrate the beekeeper.*
- *It is no good trying any of the above methods unless you ensure plenty of room for the queen to lay eggs by keeping a good eye on the state of the brood nest and, secondly, by ensuring plenty of room in the supers for honey storage. Bees like room (see Delaplane, 1997).*

SECTION 19:
EQUALISING COLONIES

The purpose of equalising colonies is to help guide all colonies to build up in a similar manner during the season and so to make apiary management simpler. It can, however, also be used to help prevent swarming in large and very populous colonies by reducing the numbers of bees in these and by using these colonies to boost the numbers in smaller colonies.

There are two aspects to equalising colonies, as follows. Equalising by:

- moving frames between colonies; and
- swapping the positions of colonies.

Procedure

If a colony is found to be weak:

1. Check to make sure that the colony is not weak because of a failing queen or disease.
2. If all is well, find the queen so that you do not transfer her and put her on a frame in a safe place in the hive away from other frames.
3. Then, either transfer frames of brood from the strong to the weak colony or swap the positions of the strong and weak colonies. Foragers arriving with nectar at a strange hive are invariably welcomed and will soon become members of that colony.

Advantages

- This is very simple – no additional equipment is needed.

- It boosts weak colonies without harming strong colonies.
- Equalising can be carried out prior to reversing hive bodies (see Section 20).

Disadvantages

- It can cause chilled brood if care is not taken (see Section 45).
- It can spread disease without you realising it. You must ensure that the colony that is providing frames has no disease. If you have swapped the positions of the colonies, make sure that neither has disease.

Pointers

- Both methods may be used to lessen the incidence of swarming by reducing congestion in populous hives.
- Both methods are most effectively carried out in the spring.
- For the best results in an apiary, colonies should be equalised from the nuc stage upwards. Keep them all the same – this makes for easier apiary management and it enables the beekeeper to compare individual colony performance better. It also enables you to make better decisions about re-culling queens, breeding and so on.

Notes:

- *Equalising colonies that are weak due to disease or a failing queen is a waste of time and bees.*
- *Equalising colonies can result in higher honey yields.*
- *Because the queen needs to be found in all cases where frames are swapped between hives, an annual re-queening can take place in conjunction with this manipulation.*

SECTION 20:
REVERSING HIVE BODIES

In the early spring, reversing hive bodies can be an effective method of swarm control, and this is one of the first and easiest manipulations you should carry out. The manipulation consists simply of swapping the positions of the upper and lower brood chambers. Often, after the winter period, the lower chamber will be empty. Placing this above the upper chamber will give the bees more room and the queen more room to lay. Bees tend to work upwards and so often the lower chamber will be found to be empty.

Procedure

1. When queen cups are started in some numbers, the first reversal should take place.
2. If cold weather prevails, reduce the hive entrance.
3. Check after about 10–14 days and, if necessary, carry out a second reversal. (You are checking to see if the colony has moved into the upper box.)
4. Reduce the entrance if required for temperature control.

Repeat the above, as needed. Usually, only two reversals are necessary. Supers can be added throughout the procedure. A queen excluder may be used above the two brood boxes. This is a very useful manipulation and should be carried out with other swarm-prevention measures.

Advantages

- No extra equipment is needed.
- It is very simple – whole boxes are moved, not individual frames.
- It is effective, especially if used as part of an integrated system of swarm control.

Disadvantage

- It is only suitable for two-brood-chamber hives.

Notes:

- *If the bees start queen cups this is not necessarily an indication of swarming or even progressing to the queen cell stage, but it can be a useful indicator.*
- *Look at other factors as well, such as hive crowding and lack of honey storage room in the brood boxes.*

SECTION 21:

DIVIDING COLONIES AND MAKING UP NUCLEI

This has many advantages for the beekeeper and is an essential component of good colony management.

Purpose

- As a way of reducing swarming, dividing colonies immediately reduces the number of bees in a colony and so alleviates the swarming pressure.
- As a way of making an increase. If you want to increase your stocks, this is an excellent method.
- For queen-rearing purposes (see Section 26).

Procedure

1. Select strong, disease-free colonies of at least eight brood frames for splitting.
2. Decide how many nuclei you wish to make up and work out how many colonies you will need to supply sufficient frames of brood (see below).
3. Find the queen and place her on a frame in a safe place. You can temporarily use a nuc box for this.
4. In each nucleus box, place at least one frame of brood and two frames of stores (or one frame and a feeder) – one frame to include pollen.
5. Shake extra nurse bees from brood frames into each nucleus box and close the lid. (Remember: these extra bees should be shaken from brood frames.)

6. Place each nucleus box in the desired location (normally at another location more than 1 mile away).
7. If the nuclei are placed in the same apiary, plug the entrances with grass and let the bees chew it away, or release them after two days. (Ensure good ventilation or place temporarily in a cool, dark room.)
8. After two days, introduce a queen cell, placing it in the centre of a frame of brood hanging downwards. Press the base of the cell as far into the frame of comb as possible so that it won't fall off the frame or introduce a caged queen (Photograph 12) (see Section 28). Ensure that the cage's entrance is facing slightly upwards and is free from obstruction. Also ensure that any plastic entrance cover has been removed.
9. The original queen may be left in a nuc on the original site, with some brood, stores, bees and empty comb or foundation.

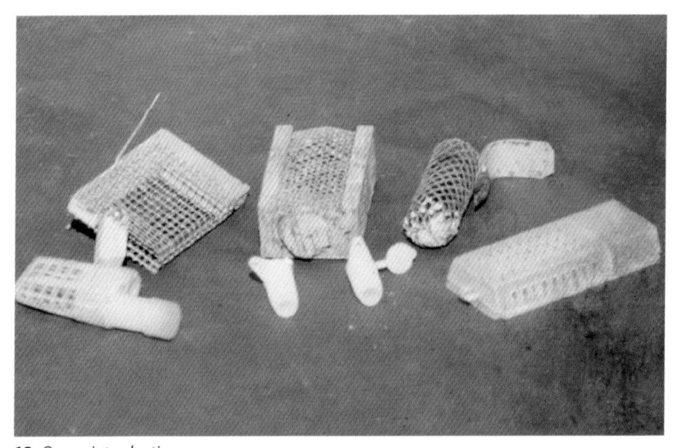

12. *Queen introduction cages*

Pointers

- Nuclei are susceptible to overheating in hot weather and may abscond, even abandoning brood.
- Nuclei are also vulnerable to robbing, so keep the entrances to a minimum or move the nuc to a site without other colonies.
- Avoid placing frames full of uncapped brood in nuclei. There may be too few nurse bees to care for them, resulting in chilled brood (see Section 45).
- Syrup feeding and capped brood help prevent absconding.

Notes:

- *The above completely splits the original colony. If this is not the intention, the colony can be split into two (see Section 22 below) or one or two nuclei can be taken from the colony, according to the same principles outlined above.*
- *There are many variations on the theme. The above method is reliable (see Johansson and Johansson, 1970).*
- *Another way of reducing the tendency of bees from the nucs returning to the original site is to place the nucs in a circle with the entrances facing the original hive. In this case, the entrances may be left open at all times. (They will still have to be moved sometime, though, as they grow into full hives.) (see Cook, 1986.)*

SECTION 22:

THE ARTIFICIAL SWARM

The artificial swarm is the classic method both of preventing swarming in a colony and of making an increase if you want to (although you can always unite the two new colonies if you don't want to increase your stock). Most beekeepers have their own variations on this method but the manipulations below are fairly standard.

Procedure *(see Figure 1)*

If queen cells are found with eggs or larvae in them in a populous colony, carry out the following procedure:

1. Move the entire hive to a new position.

2. Place a new brood box with floor in the old position.

3. Place the queen on a frame of brood in the new box.

4. Fill up the new box with new frames of comb/foundation.

5. Place the original supers plus or minus a queen excluder on the new hive.

6. Place the old hive anywhere in the apiary. The foragers will fly to the new hive on the original site.

Then:

7. Cut out all the queen cells in the old hive.

8. One week later, cut out all the new cells in the old hive, but leave one open queen cell or let the bees choose. If you don't want another hive of bees, wait a month or two and unite the two colonies (see Section 15).

Figure 1. The artificial swarm

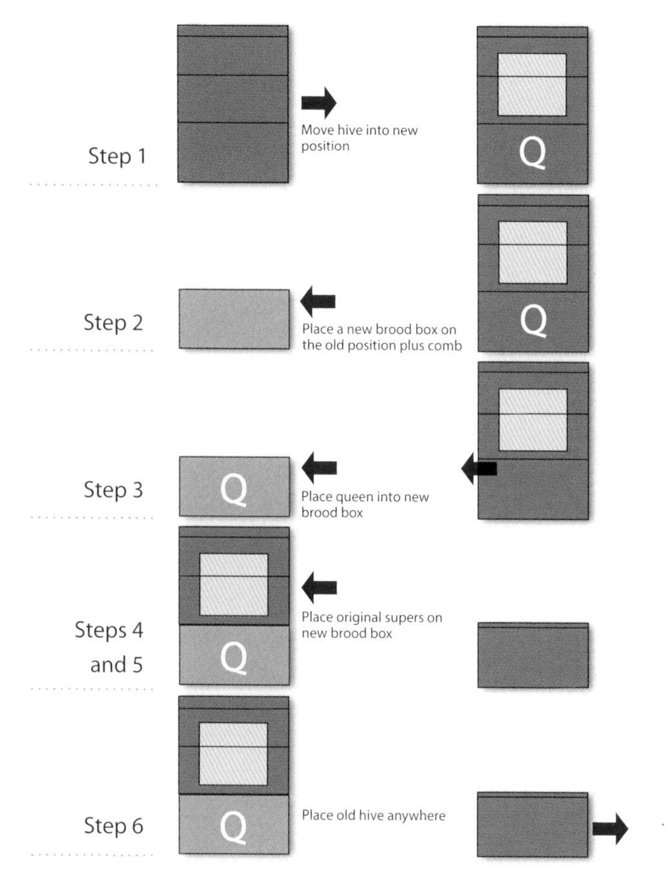

Advantages

- This is effective and is easier than it sounds.
- This method can be used to make an increase if you wish to build up colony numbers.

Disadvantages

- It is time-consuming.
- It effectively splits the colony into two smaller ones, which are thus unable to take full advantage of an early honey flow, but they can be united after everything has settled down (see Section 15).
- It reduces honey yield for that year (as does any swarming, artificial or not), but at least it's controlled.

The old hive (or, in the case of a split, one of the two new hives) will produce a new queen from the queen cell. Check this regularly until eggs/brood are seen in a good pattern, then you will be sure that she has mated satisfactorily. If not, unite or disband the colony.

If you can't find the queen, carry out the following manipulation:

1. Cut out all the queen cells.
2. Split the colony into two, ensuring that each half has eggs and/or very young brood.
3. After 3 days, look at each half. The one with eggs will be queen-right but it probably won't have any queen cells. The half with no eggs will be queenless and will probably have queen cells.
4. In the queenless colony, cut out all the queen cells except one.

This has the advantage of being less time-consuming.

Note:
If the colony is too aggressive even to contemplate finding the queen (and this happens), see Section 14 for advice.

SECTION 23:
THE DEMAREE METHOD OF SWARM CONTROL

This method of swarm control is quite drastic but usually effective. It uses the same principle as the artificial swarm method but enables the hive to be kept as one unit.

Procedure *(see Figure 2)*

1. Destroy all the queen cells. Don't miss any.
2. Place all frames of brood into a new brood chamber(s) in an adjacent position.
3. Place empty frames of comb in the original brood chamber.
4. Find and place the queen in this empty brood chamber (she will probably be with the brood in the new brood chamber).
5. The queen's new brood chamber will now be on the bottom.
6. Place a queen excluder or a super of honey above this.
7. Place a new brood chamber(s) above the excluder/honey super.
8. After 7–8 days, destroy all queen cells in the upper brood chambers.

Figure 2. The Demaree method

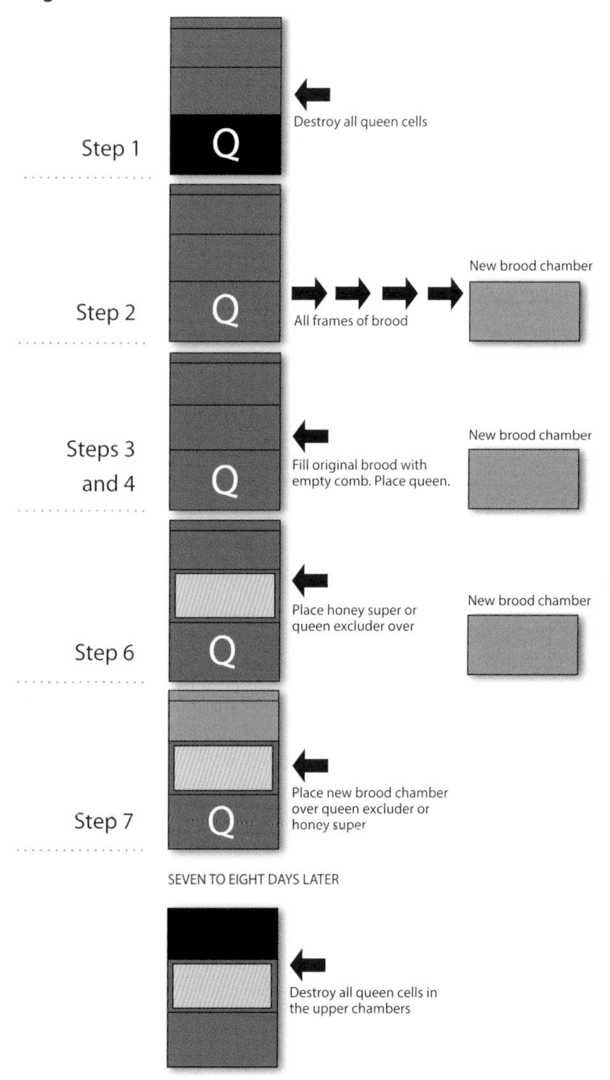

Advantages

- If carried out correctly, this is a very effective method of preventing swarming.
- It enables the beekeeper to keep the hive as one unit so that you don't have the problem of increase unless you want it.
- It will provide a very strong colony able to take advantage of an approaching honey flow.

Disadvantages

- The process is very time-consuming because of the need to find all the queen cells.
- If not followed correctly, or if you miss a queen cell (an easy thing to do in a large colony), the colony will probably still swarm.

SECTION 24:
THE QUEEN REMOVAL METHOD OF SWARM CONTROL

Like the Demaree method, the queen removal method is effective but very time-consuming, and so for this reason it is difficult to practise on a large scale. Brother Adam used this method on a regular basis and claimed there was no doubt about whether it would work or not (see Brother Adam, 1975). Most commercial beekeepers who hold large stocks wouldn't use this method because of the time it takes.

When the colony is found to be producing queen cells, carry out the following procedure:

1. Find and remove the queen.
2. Destroy all the queen cells except one (select this carefully), *or* destroy all the queen cells and insert one of your own, *or* destroy all the queen cells and introduce a new queen two weeks later, *or* reintroduce the original queen you have kept safe in a nucleus box.
3. Seven days after steps 1 and 2, destroy any new queen cells.

Advantages

- This is simple and reliable.
- It can be used in conjunction with re-queening because you have to find the queen anyway.
- No extra equipment is needed.

Disadvantages

- It is time-consuming because of the need to find all the queen cells.
- The time between queen removal and a new queen laying eggs can be 3 weeks or more and, during this time, the colony may do little work even if there is a honey flow on.

Note:
If, after finding and removing the queen at step 1, a virgin is about to emerge, she can be left on the comb. The colony will probably not swarm.

SECTION 25:
BAIT HIVES AND SWARM COLLECTING

During the swarming season it is a good idea to collect swarms by letting the authorities (usually the local council) know that you will do so and/or by setting up bait hives to entice other people's livestock into legally becoming yours for free!

Procedure

1. Make up a bait box using an old hive body (full-depth Langstroth size recommended), although it is best to use a hive body that you use in your apiary. (Ensure it is free from disease.)
2. Fit on a floor without an entrance and make a lid.
3. Place in the box two old combs that are free of disease and another two of foundation.
4. Close up the box and drill a hole 10 mm in diameter near the bottom of the front of the box.
5. Place the bait hive 1m from the ground away from direct sunlight and where it will be sheltered from high winds.
6. If a swarm enters, place it in a hive where required when the bees have ceased flying, and replace the bait hive in its position.
7. After a couple of days, treat the swarm for Varroa.

Advantages

- There is little, if any, expense in setting it up – use old kit and just sit and wait.

Disadvantages

- If the box is left for too long, wax moth may become a problem.
- If the weather is hot, the foundation may buckle (unless plastic foundation is used).

Variation

Use foundation only and a pheromone lure (Nasonov pheromone). This is more expensive than the previous method. Commercial swarm lures are available on the market.

Pointers *(see also Schmidt, 2001)*

- Propolis and other hive remnant odours are a powerful attractant to scout bees (this is a likely scenario in nature).
- Swarms will generally prefer hives containing propolis over those that do not.
- Swarms prefer hives containing old comb over hives with hive remnant odours but no old comb.
- If using a Nasonov pheromone lure, swarms are attracted solely on the basis of the lure. Old comb and hive remnant odours are not important.

Note:
Research shows that odours from substances not of bee origin are neither attractive nor repellent to swarms. The same has been shown of odours from bee diseases. The research therefore shows that beekeepers using other attractant substances are basing their swarm attracting more on luck than judgement. Many of these mixtures are sold by bee-supply shops, so avoid them and save money. Old comb is usually free.

Author's note:

1. *My Spanish bees were good workers but could get out of hand and swarmed incessantly* (Apis mellifera iberica). *Until recently, beekeeping practice in Spain encouraged the swarming tendency for increase.*

Part E
Queen Bees

Introduction

There are many advantages to rearing your own queens, even if you only use the simple methods. You have a degree of control over timing and knowing exactly when you will have queens for re-queening purposes, and you will also have a certain amount of control over the selection of some favourable traits – assuming you don't rear new queens from bad colonies! You may even decide to sell a few surplus queens and start your own business in queen rearing. There are never enough queen rearers and new competition is always welcome.

You can rear queens with the minimum of equipment – in fact, with no equipment whatsoever if you just let the bees do it all for you (not recommended but it happens to most of us). If, however, you want to rear your own queens you must start off by deciding exactly which method you are going to use and which method of larval presentation you are going to use (see Figure 3). There are three basic methods of presenting larvae to nurse bees to raise into queens. You can:

- allow the bees to decide and let them raise the queen cells to the stage where you cut them out for use. They introduce the larvae to the nurse bees;
- provide the queen with comb and suggest to the bees the best position to build queen cells, as in the Miller method (see below); or
- inspect young larvae laid by the queen, choose which ones you want and transfer them to artificial queen cups ready for the nurse bees to finish (larval transfer or grafting).

This Field Guide will take you through two non-grafting methods and one grafting method. All you need to do is ensure that you have the required equipment ready. For the beekeeper out in the field, the steps will act as a valuable aide-memoire in what can be a complicated series of manipulations.

Figure 3. Methods of larval presentation

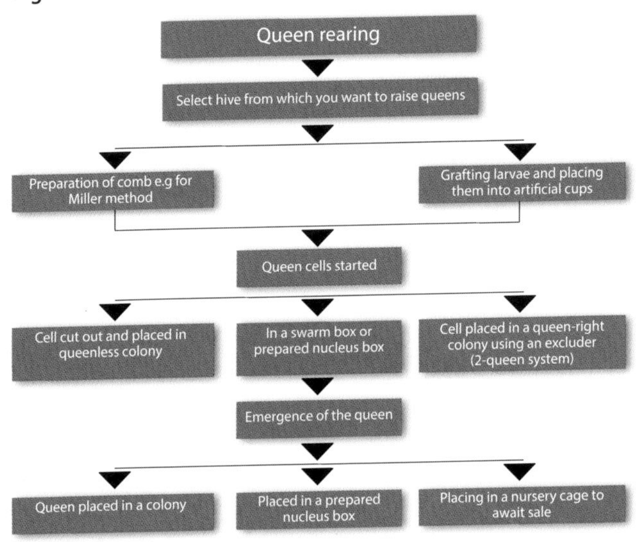

There are two very important things to remember in rearing queens:

- Timing is all important. Once you have started, you can't stop and you have to attend to each stage at the correct time.
- You must keep a record of what you are doing so that you don't miss a stage or duplicate a stage.

Figure 3, which is loosely based on a diagram first produced by E.B. Wedmore (a former master beekeeper), should help you understand the different stages of queen rearing.

As you can see, queen rearing is in four parts, as follows:

1. Preparing larvae for presentation to queen-rearing units. This can include letting the bees choose their own larvae, grafting small larvae of about 24 hours old from your chosen colony on to either plastic or homemade wax cells, or using one of the queen-rearing kits whereby the queen is trapped in a cage on a comb and is able to lay in prepared plastic cells. These can later be removed and presented to a queen-rearing unit.

2. Preparing queen-rearing units to receive the selected larvae. These units are strong colonies in which the queen has been removed or is separated from the queen-rearing part of the hive by a queen excluder.

3. Transferring the resulting queens/queen cells to receiving hives/nuclei from where they will be allowed to fly and mate.

4. Finally, the mated queen(s) will be placed in a queenless nucleus or hive ready to begin work as the queen of a colony.

Pointers

- Whatever method you decide to use, the colony that is to rear queens (or the part of the colony that is to rear queens) must be queenless. De-queen thoroughly and then check to see if there is another queen lurking about. This does occur.

- Treat eggs and very young larvae carefully. Neither should be exposed to too much direct sunlight.

- Use only healthy colonies. Make very sure of this.

- Timing is all in queen rearing. Failure to keep accurate records can be expensive and disastrous.

SECTION 26:
FIELD QUEEN REARING

Let-alone method – variation 1

This variation is for those beekeepers who wish to rear queens quickly and efficiently while out in the field but who have no queen-rearing equipment readily available. The 'let alone' methods are based on the fact that queenless bees will try to raise a queen from young larvae. With this method you can:

- rear queens from a chosen stock (a stock with a productive, gentle queen);
- keep control of the number of new queens you want to rear; and
- complete the queen rearing to your own timetable.

Pointers

This method (especially the basic method) corresponds most closely to the way colonies reproduce in nature. A surprisingly large number of beekeepers rely on this method both by accident and by design.

Additional equipment needed

A sharp knife or scalpel and the required number of prepared nucleus boxes. (Nucs with a frame or two of brood and bees, and a frame of stores as minimum.)

Procedure

1. During the spring, check that the colony has plenty of flying drones.
2. Prepare a nucleus hive (see Section 21).
3. Check to make sure that the colony has plenty of young larvae and eggs and then remove the frame with the queen on it and place in the nucleus hive. (This is to keep her safe and well in case your plans go wrong.)
4. Inspect the colony 48 hours later to see if queen cells have been started.
5. Carefully cut out the queen cells 10 days later and place each in a prepared nucleus box (see Section 21).
6. Check the nuclei after 2 weeks to ensure that the queen has emerged, mated and is laying eggs.
7. The nucs can then be placed on destination hives ready for transfer to these hives.
8. Now you can decide whether you require the original mother queen still in her nuc. If she is a productive and gentle queen, you may wish to rear from her again.

Advantages

- This method is easy. The bees do all the 'thinking'. The beekeeper can usually assess and choose the best cells from a good selection, and very little extra equipment is needed.
- The resulting queens may also be used to re-queen hives (see Section 28) at your annual re-queening session.

Disadvantages

- The beekeeper has no control over the timing of queen cell production and assessing the age of the cells can be difficult. Remember that, if cells are cut out and moved too early, the queen may be damaged.
- The bees raise queens laid in worker cells, not in specially prepared queen cups. This may mean they choose larvae that are not of optimal age, thus affecting the resulting queen quality.

Note:
You will note that, with this method, if you find sealed queen cells then it is likely the colony has swarmed already, so look for the open queen cell and for a virgin or young queen. The bees may have re-queened themselves for you.

Let-alone method – variation 2

1. In the spring, as soon as drone larvae are plentiful or drones are flying, split a colony in two, ensuring that each half has both very young larvae and eggs. (If possible, move one half to a different apiary.)
2. Two days later, check both halves. One half, the half without the queen, should now have queen cells of a known age.
3. Ten days after splitting the hives, assess the cells and cut out those you find suitable for placing in mating nuclei.
4. Leave two good queen cells in the still queenless part and keep it separate or destroy them and reunite the colony.

Advantages

This method has two advantages, as follows:

- It can be used to produce queens and as a swarm-control method. There is no requirement to find the queen although, after day two, you will know which half she is in.
- This method can also be used to make an increase in your stocks if needed.

Disadvantages

Although not essential, another apiary site is an advantage. More work, time and care are involved. It is important to note that, by using the two variations, the beekeeper is causing the colony to raise queens from what were originally worker cells under 'emergency' conditions. Research has shown that queens produced in this way may, on average, be inferior.

The Miller method *(larval transfer not necessary)*

This is a very easy and more controllable method of rearing queens. It gives you a greater degree of choice in which larvae to choose.

Procedure

1. Take a frame of foundation and trim this to form triangular shapes, as shown in Photograph 13.
2. Place this frame into a strong nucleus hive which contains a queen of your choice and lots of bees and frames of brood and honey.

13. *Miller frame. Cut the foundation like this. The bees will draw out the wax and they will develop queen cells along the angled edges*

3. Shake in several frames of bees. Shake the bees off the brood combs so that you get the nurse bees and shake to overflowing. The queen will have very little room to lay eggs and the bees will quickly draw out the wax of the Miller frame into cells. The queen will then lay eggs in the frame.

4. *Or* place the Miller frame into the middle of the brood box of the hive in which the favoured queen lives. (Whichever way you do it, you are basically enticing the queen you want progeny from to lay the eggs.) So far, you have control of the bee you want and of the timing of everything. After 6 days, prepare a good, full nucleus of bees with stores and emerging brood but no queen.

5. *Or* de-queen a strong colony you were going to re-queen anyway.

6. *Or* place a queen excluder over a strong colony's brood box (ensuring that the queen is below the excluder). Transfer to the top box a couple of frames of emerging brood and some stores. (Make sure your queen excluder is

sound.) Again, whichever way you do it, you are ensuring that the colony that will develop your queen cells will be queenless for one day at least before introducing the cells.

7. One day later, go back to your Miller frame and take a look at the edges of the triangles of comb. The larvae along the edge of the triangles aren't in fact larvae but eggs.

8. Trim these edges back to where you can see the tiny comma-shaped larvae that will be about 24 hours old or younger. They really will look like a tiny c (don't confuse them with eggs). Each cell along these edges will contain just the right type of larvae once you've trimmed the eggs away.

9. Following the edges of the triangles, destroy with a matchstick every two larvae, leaving the third to develop.

10. Place the Miller frame into your queenless hive/nuc that has been queenless for a day. (Check there are no queen cells that you may have missed and no open brood in this hive/nuc. If there are, the bees may not accept your Miller frame larvae.) The bees will now develop the larvae along the edges of the trimmed frame in the queenless hive. They will automatically draw out queen cells along the bottom edges and the slanting edges.

11. After 10 days, these cells can be used for re-queening purpose. Just very carefully cut them out with a sharp scalpel ensuring you don't press or squeeze them and that you don't cut into the cells.

Summary

- **Day 1:** make the Miller frame from good foundation wax. Place it into a full and queenless nuc/hive or a hive with a queen excluder.
- **Day 6:** prepare your queenless cell-building nucs or queenless hive.
- **Day 7:** remove your Miller frame, trim the edges back to the little c-shaped 18–24-hour-old larvae and place the frame into the centre of the queenless hive/nuc which will act as the developer hive.
- **Day 16:** if you are going to go a step further and rear mated queens, prepare your queenless nucs, each one ready to receive one queen cell. Remove them to another apiary, block their entrances with grass or place them in a circle around the original hive from which you obtained the frames to make up the nucs.
- **Day 17:** carefully cut the developed queen cells from the Miller frame and use them to re-queen your hives, sell them or place them in the prepared mating nucs.

Advantages

- This is a low-tech method that gives you very good control over timing and quality.
- You can keep repeating it by reinserting your Miller frame.

The swarm-box method *(larval transfer necessary; see procedure below)*

This method is probably the most used queen-rearing method, especially by serious queen rearers. It is not applicable to the let-alone method of larval presentation because you need either to graft some larvae yourself or to use one of the queen-rearing kits, such as the Jenter or Cuckpit systems, to prepare the larvae (available from all bee-supply stores). This method is suitable for continuous production.

Additional equipment needed

- A well ventilated, three or four-frame nucleus box.
- A frame feeder.

Procedure

1. Take a well ventilated, three or four-frame nucleus box. Into this shake bees off three well covered combs of brood. If you shake off brood frames you will be more certain of shaking off the nurse bees you want.
2. Into the box place a full frame of mainly honey and some pollen, and a full frame of pollen. No brood. Leave a gap between the frames.
3. If the box takes four frames, put in a frame feeder full of sugar syrup.
4. Prepare the larvae for acceptance by your chosen method and place this frame into the box between the two frames of stores.
5. Close up the nuc and place it in a dark, cool room for about 24 hours. Note this in your diary.

6. About one hour before opening up the nuc, take a populous and healthy colony of at least two storeys and confine the queen to the lower chamber with a queen excluder.
7. Place into the centre of the upper chamber a frame of young larvae (this attracts nurse bees) and a frame of pollen. Make a gap between them.
8. Fill the rest of the chamber with sealed brood or stores.
9. An hour or two later, open up the swarm box and transfer the started queen cells to the main colony above the queen excluder and close up. Note this in your diary.
10. The cells stay there until they are transferred to the mating nuclei – i.e. just before the virgin queens are about to emerge. (Meanwhile, another batch of prepared larvae can be placed into the swarm box.)

Advantages

- This is fast and simple.
- It can be used for continuous production.
- It is a very reliable method.

Disadvantages

- It involves a transfer during the proceedings.
- It involves two hives (a nuc and a full hive).
- It is not really the best method to use if you only require up to, say, 10–12 queens.

Note:
If you are unable to transfer the started queen cells to a builder hive within the time frame outlined above, don't worry. Just make sure that you place the swarm box outside in a convenient position and let the bees fly. Transfer the frame of cells when you can. You must, though, remove the queen cells to mating nuclei (or an incubator) 10 days after introduction. Swarm boxes are not designed for this but, in extremis, *they will produce queens.*

Larval transfer in queen rearing

The term 'larval transfer' describes accurately this method of rearing queens, but the term 'grafting' is used colloquially. To transfer small larvae from cell to cell requires excellent close-up vision. For many beekeepers, reading glasses and/or a magnifying sheet may be necessary. If all this looks a bit complicated, look again. It really is simple, invariably successful and well worth the effort.

Extra equipment needed

- A bain-marie to melt wax in safely. Be careful when heating wax – it can explode.
- Some wax, new or recycled.
- A wooden cell former (a piece of 8 mm dowel with a rounded end).
- A frame with a cell bar(s).
- A 'grafting' tool. There are several different types of grafting tool and most are available from bee-supply firms. I prefer to use the Chinese grafting tool because I find it easier to pick up and deposit larvae without rolling them.

- A small, sharp scalpel.
- A magnifying lens if required.

When you first start this process, you need to prepare certain items in advance, so the procedure starts off with a preparation phase.

Preparation phase

Prepare a frame fitted with two cell bars. Make some artificial queen cells, as follows:

1. Round off an 8 mm diameter piece of doweling as a cell former.
2. Make a mark on the doweling 5 mm from the rounded end.
3. Place the rounded end in a glass of water for an hour or so.
4. Melt some wax in a bain-marie and remove from the heat.
5. Take the dowel rod from the water, shake the water off and briefly dip it into the melted wax up to the mark.
6. Remove from the wax and plunge into the water.
7. Repeat this process about five or six times.
8. Carefully rotate the wax off the dowel. You have a queen cup.
9. Make about 30 of these.
10. Run some molten wax along the underside of the cell bar(s).

11. Place a blob of molten wax 4 mm along a cell bar and, before it dries, place a queen cup on this. If you are using Langstroth frames, you can place 12 cups on each cell bar. If you use British Standard frames, then place 8–10. The more you space them out, the easier it is to remove them. You now have your queen cells prepared.

Or having prepared your cell bars, place plastic queen cells with their special mounts on to the bars (see Photograph 14). There is no difference in acceptance rates between wax or plastic queen cells, but plastic cells are awkward to clean afterwards. You simply re-melt wax ones.

14. *Cell bars in frames with the plastic cells (queen cups) hanging*

Procedure for moving the larvae

1. Choose a colony from which you want to raise queens.

2. Place a frame of new comb (not foundation) into the centre of the brood nest.

3. Eight days later, you will find eggs and newly hatched larvae ready for transfer. These small larvae will usually be situated around the edges of older larvae and capped cells. From this colony take the comb of eggs and very young larvae. (It makes it easier if the comb is of new or nearly new wax – not old black stuff.)

4. Now take this frame to your shed or car/lorry cab and inspect the larvae carefully. Look for those that are so small you can hardly see them. The ones you want will resemble a small letter c or comma sitting in the bottom of the cell in a bed of royal jelly. Use a magnifying sheet if necessary.

5. With a scalpel, pare down the cell walls of a row of such larvae. This makes it easier to remove them. Using your favourite grafting tool (I prefer the Chinese grafting tool – in fact, it's the only one I can use successfully), transfer each chosen larva from its original cell to the artificial cell (either wax or plastic). If you roll a larva or in any way damage it, discard it.

6. When you have transferred the required number of larvae, place the cell bar(s) into the down position and put the cell frame into an empty nucleus hive for safety. Return the rest of the brood frame to its brood chamber. Remember that, if you produce too many queen cells, you may not have

enough colonies to provide bees to make up the necessary number of nucs. If you need 20 queens, go for 30 queen cells to allow for failures and wastage.

7. Place the cell bar(s) into the down position and put the frame into the type of rearing colony you previously decided upon. Some 48 hours later, check the bars for acceptance. You will know if the cells have been accepted because the bees will have drawn out the cell and the larva will be floating in a bed of royal jelly.

8. You will now know how many nucs to prepare for the number of queen cells.

9. Ten days after grafting, remove the queen cells with a scalpel (or unplug the plastic cells) and place into a prepared mating nucleus.

SECTION 27:

QUEEN-CELL TROUBLESHOOTING GUIDE

(see Table 10)

During your queen-rearing activities, whether using grafted larvae and cell bars or one of the non-grafting methods, problems may occur. This section deals specifically with the problems affecting ripe queen cells in the rearing box or nuc, whether on a cell bar or on the comb.

TABLE 10: Queen-cell troubleshooting guide

Problem	Cause	Solution
Queen cell found on the floor of the mating nuc	The cell was not securely attached to the brood comb	When you cut out the cell from its original hive, make sure that you cut a large flange of wax as a base for the cell. Press this flange securely into the brood comb of the nuc and add brace comb wax if necessary. Within a day, the bees will make it secure, so it's important that it's securely fastened to last for 24 hours
Queen cell ripped apart prior to the queen's emergence	If the cell is placed in the brood comb and it extends too far above the comb's surface, it can be joined with brace comb to the next frame by the bees	Make sure that, when you initially attach the cell, you push it as far as possible into the comb so that it hangs downwards and doesn't protrude too far
A queen cell on a bar is empty even though it appears normal	Often you will find that, if this is the case, other cells will have been partially ripped open.	Ensure that your timing has been right. You can protect cells with cell protectors available from

	This is because a virgin has emerged and has destroyed the other cells. She has ignored the apparently healthy cell because the queen inside has died of other causes	bee-supply stores (hair curlers can be used for this)
No queen emerges from an apparently normal cell	When you inspect the cell you find it to be empty. This can mean that the queen has emerged and the bees have resealed the cell	Look for the queen and/or eggs to verify this
The queen cells on a bar have holes in their sides	The cell occupants are dead. Either a virgin has emerged earlier than the rest and killed her rivals,or a queen or virgin has passed through a queen excluder and entered the queen-rearing part of the hive	Make sure there are no queen cells in the excluded part of the hive. Use sound queen excluders and cell protectors. Always inspect your queen excluders carefully before use

Assessing queen cells

Always try to choose good queen cells. Cells can vary due to their developmental environment. Cell size and appearance may differ greatly. Highly sculptured cells receive more attention from bees than smooth cells, and highly sculptured cells and a larger size generally go together (see Photograph 15).

15. *A well sculptured queen cell*

Part E

SECTION 28:
RE-QUEENING AND ASSESSING QUEEN CELLS

Re-queening can be carried out in the spring and the autumn/fall. There are advantages to both.

Practical requirements

For the hobbyist, this depends on:

- your mindset;
- very much on your family schedule;
- your integrated pest-management strategy; and
- your anti-swarming strategy.

Spring re-queening: advantages

- Re-queening in the spring can be allied to a swarm-control strategy.
- Re-queening in the spring can be allied to an apiary stock-increase strategy.
- For many hobbyists, this is the start of the active season and they can watch their new queens and colonies grow.

Autumn/fall re-queening: advantages

Replacement in the autumn/fall means that:

- young queens are less likely to die out over the winter;
- queens generally cost less if buying from a supplier;
- the population growth of tracheal mite can be disrupted;
- the population growth of Varroa can be disrupted; and

- there is no break in the brood cycle in the spring when the colony should be rapidly building up for honey production.

Research that backs this up

- US research shows that tracheal mite build-ups are best disrupted at this time.
- Australian research shows that new queens are less likely to die out over winter and that more spring queens overall are rejected by the workers than autumn/fall queens.
- There are generally more drones around in the early autumn.
- Generally, there are more drones available to mate with queens in the early autumn than in the spring and so, if you are rearing your own for queen replacement, you will have a better chance of success.

Assessing queens

The queen will fly and mate 5–6 days after emergence and will start to lay eggs 36 hours or more after her mating flight. Therefore check the nuc or hive two weeks after placing queen cells, and look for the following:

- On one frame at least, a small, round patch of sealed brood surrounded by uncapped brood. The brood should become younger towards the edges of the patch.
- There should not be too many empty cells in the area of capped brood.
- Eggs around the edges of the unsealed brood.
- The location of the queen to ensure she is alive and well.

SECTION 29:
MARKING QUEENS

Queen bees are marked on the thorax for various reasons (see below). No research evidence shows that marked queens live shorter lives or are superseded more quickly by workers or produce fewer eggs than unmarked queens, unless they are damaged by handling in the marking process.

Advantages

- A marked queen is more easily identified.
- If you mark a queen and later find an unmarked queen, you will know that either swarming or supersedure has taken place or that your marked queen has died.
- Using the International Marking Code (see Table 11), you can tell the queen's age.
- Marking queens with coloured discs or numbers can be of use in research and is helpful for identifying specific strains, lineages or other qualities.

Disadvantages

- You may damage the queen while marking her, which could lead to her rejection by the colony or a reduced egg-laying rate or even to a drone-laying queen.
- You may become accustomed to looking for dots and not queens and so may miss a queen without a mark. You could then assume a queenless colony and make the wrong decisions concerning that colony.

Methods

1. The queen may be picked up manually by placing the fingers on the sides of the thorax or she can be held by holding one of her hind legs. Once picked up, mark her thorax, allow the paint or glue to dry and place her back in the colony. This takes some practice and is the method most likely to damage the queen. Try it out on drones first. If you try it on workers you will be stung. The cage method (see below) is better for the queen. There should be no need to handle her at all.

2. A small cage can be constructed or purchased which is held over the queen on the comb. The workers can escape through the cage but the queen is too big. The cage is pressed on to the comb until the queen is trapped fast and the paint is then applied. Wait until the paint is dry before removing the cage. This method is simple and is less likely to damage the queen.

3. A queen catcher may be employed to put the queen into a marking cage whereby she is pushed up to a screen by the use of a sponge plunger. Once trapped against the screen, paint is applied to the thorax and allowed to dry.

TABLE 11: The International Marking Code

Year	Colour
0 or 5	Blue
1 or 6	White
2 or 7	Yellow
3 or 8	Red
4 or 9	Green

Thus a queen for 2010, for example, will be colour marked blue.

SECTION 30:
INDUCING SUPERSEDURE

There is plenty of research to show that supersedure can be induced by 'tricking' the bees into thinking that a virgin emerging from a queen cell placed there by you has come from a supersedure cell, and so they will accept it as such.

Why attempt supersedure?

By encouraging the bees to accept a supersedure cell you will be able to replace a queen without the swarming impulse rearing its head.

The following is the procedure for inducing supersedure:

1. Place a ripe, protected queen cell from the chosen stock between two frames of a honey super.
2. Place a queen excluder between the brood box with the old queen and the honey super with the queen cell.
3. Make a small entrance in the super box.
4. Leave alone until the resulting queen can be seen laying eggs.
5. Mark her to distinguish her from the old queen.
6. Remove the queen excluder.
7. Wait to see if she is accepted by the colony.
8. Smoke the queen and bees down into the brood box and replace the queen excluder.

Note:
Position may be an important factor in placing the supersedure cell because many beekeepers say that a cell in the middle or upper part of the comb is more likely to be a supersedure cell.

Pointer

The cell should be placed in a honey super, not the brood frame.

Part F
Honey harvest procedures

Introduction

For most beekeepers the harvest is the aim of the game and there are many manipulations that can help the beekeeper to maximise this. Honey can be harvested as the season progresses or at the end of the season. Much of this decision may be based on the existence of the honey flows in the area, and you must know about these flows either by asking or by experience. For example, if you have an early flow of, say, dandelion and then a later flow of thistle, you may wish to sell the two honeys separately.

Pollen and propolis harvesting are included in this part and wax rendering is also looked at. These products have a ready market, and wax is especially useful for all beekeepers, either as a side product for sale or for recycling into wax foundation sheets for use in hives. Wax blocks can also be swapped by many bee-supply companies for foundation.

SECTION 31:
MAXIMISING HONEY PRODUCTION

Bee numbers are important

A colony of 60,000 bees will out-produce two colonies of 30,000 bees simply because around 15,000 or so bees in a hive are required to look after the queen, to nurse brood and to be housekeepers. Thus two colonies of 30,000 will have a total of 30,000 bees at home and 30,000 out foraging. One colony of 60,000 bees will have 15,000 bees at home and 45,000 out foraging. So it is important to build big colonies early on to take advantage of the first flow – and not to lose bees due to swarming. A two-queen management system invariably ensures adequate numbers of bees in a hive and helps to boost honey production (see Section 32).

The following manipulations will help to ensure a good harvest and have already been dealt with in previous sections:

- Ensure that your apiary(ies) is correctly sited near to ample forage (see Sections 1 and 4).
- Spread the brood out on a regular basis (see below).
- Equalise your colonies (see Section 19).
- Prevent swarming (see Section 18).
- Unite colonies (see Section 15).

The following manipulations will also assist in maximising honey production.

Build up numbers by early feeding

- Brood production in a colony must commence at least 40–45 days prior to a honey flow. This is the time it takes from egg to forager. The early feeding of pollen or a pollen supplement will boost this production well in advance of the flow.
- Give early bee numbers a boost by early spring feeding of invert sugar syrup (see Section 57).

Adequate supering – at the right time

- If you know that your bees can produce a couple of supers of honey in a given flow, place these on beforehand plus an extra box just in case. Don't put too many boxes on because the bees have a tendency to fill the middle combs only of each box, which means a lot of extra work for the beekeeper when harvesting.
- Some beekeepers believe that under-supering is more effective than over-supering.

Re-queening in the autumn/fall

Re-queening at this time has many advantages whereas by replacing a queen during the spring build-up for the main nectar flow, the egg-laying and brood-rearing processes will be interrupted. This will, in turn, reduce the colony population and, consequently, honey production.

Provide entrances in the supers for returning foragers during a honey flow

This tactic works well in a very busy flow. The foragers can return directly to the storage site rather than fighting their way up through the brood nest. A split board placed between every other super allows this without the need to drill holes.

Notes:

- *Always try to use frames of comb for the supers. Bees don't like to build out cells during a busy flow and they use 8:1 honey to wax when doing so, thus cutting down the harvest.*
- *Try to use a swarm's propensity to build comb as much as possible, if you can capture them.*

Spreading brood

This can induce a colony (headed by a good queen and free of disease) to build up more rapidly in the spring and so be in a better position to take advantage of the honey flow.

Procedure *(see Figure 4)*

1. Move the whole brood area to one side of the hive.
2. Place a comb of stores between it and the side wall (comb 1).
3. Find the comb with most sealed brood (comb 4).
4. Place this to the right of comb 7 (the last comb of brood).
5. This induces the queen to move to comb 6 (was comb 7) in which she lays. (Being between two brood combs, it is now warmer.)
6. Some 7–10 days later, make another such shift.
7. Continue until a large, even brood pattern has been obtained.

Frames may be transferred according to the same rules without moving the brood frames to one side of the hive.

Figure 4. Spreading brood

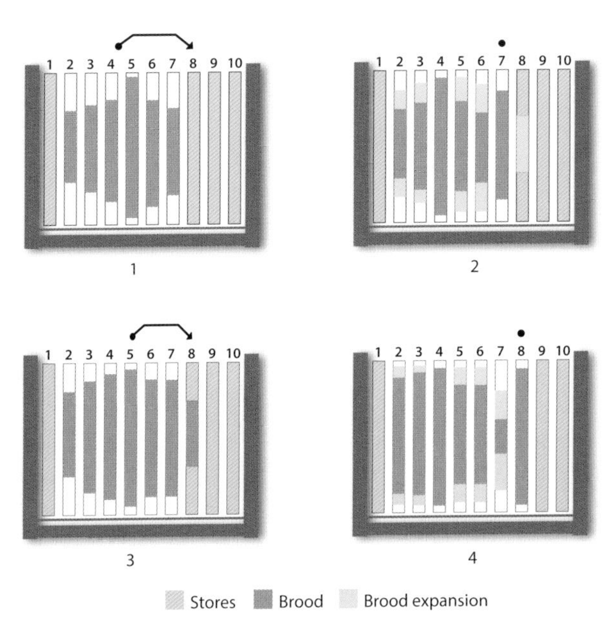

NB: The diagrams show the movements of the combs described in steps 4–6. Comb 1 is shown to be a frame of stores, put against the side wall to protect the brood from cold. In warm areas, this may not be necessary

Advantages

- This method speeds up colony growth with no input of brood frames from another colony, therefore avoiding the spread of disease.
- Brood chamber expansion proceeds in one direction, therefore making it easier to assess.
- It is a simple and effective procedure.

Disadvantage

- Great care must be taken, and so this manipulation is time-consuming.

Notes:

- *Never transfer a brood frame over an empty frame, thus isolating it. Chilled brood may result (see Section 45).*
- *Don't rush this task. It takes slow and careful manipulations to succeed.*

SECTION 32:

A TWO-QUEEN MANAGEMENT SYSTEM

The two-queen management system is used by some beekeepers to maximise honey production (see Brown, 1980; Ambrose, 1992). Experience among large-scale commercial beekeepers shows that two-queen colonies can consistently produce better honey yields than single-queen colonies but at a cost in time and workload. Below is one method of preparing a two-queen hive.

Procedure *(see Figure 5)*

Use strong, over-wintered colonies treated for Nosema (see Section 46). Spring feed with pollen/substitute and sugar. Two months before the start of the flow, divide the colony as follows:

1. Place the old queen, young brood (uncapped) and about half the bees in the bottom chamber.
2. An empty brood chamber with drawn comb may be placed above this (I don't bother to do this).
3. Cover with a division board.
4. Place a new queen with capped brood and half the bees in the upper chamber.
5. Above this place an empty brood chamber with drawn comb, if available.
6. Carry out brood chamber reversals if necessary in both the upper and lower units (see Section 20).
7. After two weeks, replace the division board with a queen excluder.

8. Super as required.
9. About 1 month before the flow ends, remove the queen excluder and combine the two colonies. The old queen is usually killed.
10. Winter the colony with the young queen.

Figure 5. Two-queen management system

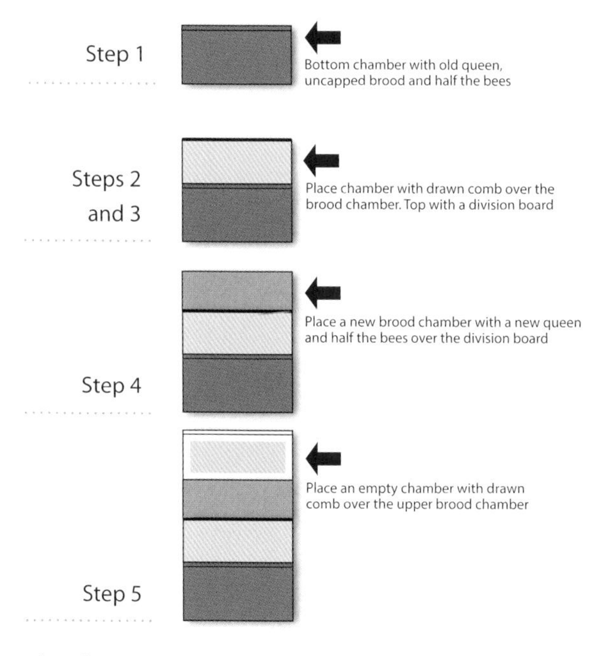

Then follow steps 6–10

NB: The upper chamber over each of the brood nests may be deep or shallow size

Advantages

- This can maximise honey production in areas where a given nectar flow can be predicted.
- It tends to reduce swarming as the brood nest is split up and because of the use of young queens.
- Colonies tend to be equalised during the set-up process.

Disadvantages

- It is labour intensive, but not too much so.
- Timing has to be near perfect to use the system to best effect.
- It is not as effective in areas of unpredictable honey flow.

SECTION 33:
HONEY HARVEST CHECKS

Honey can be harvested either crop by crop or at the end of the season, but always ensure that at least 70–75% of the cells on each frame to be extracted have been sealed over.

Essential equipment to take to the field:

- Your usual beekeeping tools.
- Bee escape boards.
- Bee brushes (if used) or a bee blower.
- An escape board (if used). The multi-valve Canadian-type clearer board is recommended for speed of access for the bees because it allows more bees to return to the brood nest at any one time.
- Fume boards (if used). These should be employed exactly as indicated by the manufacturer of the bee repellent. Use too much and you will confuse the bees, and they cling on harder to the honeycomb.
- A wheelbarrow or hand barrow (usually essential to shift heavy boxes to the car or truck – or back to the kitchen).

Essential equipment in the extraction plant/kitchen:

- A bee-proof room, preferably with hot and cold running water and an electrical supply.

- The extractor, radial or tangential. This must be made of food-grade plastic or stainless steel. Ensure it is firmly mounted on a stand especially if it is motorised.
- An uncapping knife or fork, either steam or electrically heated or just a serrated-edged bread knife.
- A large bowl of very hot water to heat the bread knife if you go that way.
- A large honey bucket made out of either food-grade plastic or stainless steel to store the honey after you have emptied it from the extractor. Fit a tap near the base if possible.
- A filter device to filter honey between the extractor and the storage bucket. (Some beekeepers use a sieve with muslin in it to provide a finer filter than a sieve alone. But the finer the filter, the slower the operation.)
- Floor trays or newspaper to keep dripping honey off the floor.
- A large container to hold the cappings. This will have to be emptied regularly into a spare bucket.

SECTION 34:

POST-HONEY HARVEST CHECKS

The post-harvest storage of honey is an important part of the harvest operation, and so the following checks should be carried out.

Moisture content

Fermentation in honey – because of excess moisture content – can ruin the crop. Fermenting honey can also blow up its containers. Use a well calibrated honey refractometer (electronic varieties now exist) to measure moisture content immediately after extraction. Use Table 12 to assess the danger of fermentation, which also depends on the yeast content.

Fermentation of stored honey due to yeast/moisture content

This natural fermentation will depend on the moisture content of the honey. Always use a calibrated hydrometer.

TABLE 12: The danger of fermentation

Moisture (%)	Liability to ferment
Less than 17.1	Safe regardless of yeast
17.1–18	Safe if yeast count <1,000/g
18.1–19	Safe if yeast count <10/g
19.1–20	Safe if yeast count <1/g
Above 20	Always in danger

Hydroxy-methyl-furfuraldehyde (HMF) checks

In most countries a level of HMF above 40 mg/kg is illegal because it may signify excessive heating during extraction. Honey stored in a warm place can accumulate a high level of HMF, so beware and note the temperatures and times given in Table 13.

TABLE 13: HMF– temperatures and times

Temperature (°C)	Time*
30	100–300 days
40	20–50 days
50	4–10 days
60	1–2.5 days
70	3–5 hours
80	<2 hours

*Time for 30 mg/kg of HMF to accumulate.

Detecting levels of HMF in your honey – a rough guide

This is a rough but easy test to give you an indication of HMF levels in your honey. Honey that has been stored for some time and is for sale should be checked. Carry out the following procedure:

1. Mix 10 g of honey with 40 ml of distilled water at 20 °C (room temperature – don't warm the liquid).
2. Leave for 1 hour keeping it at that temperature. (The glucose oxidase will give off hydrogen peroxide.)
3. Immerse a hydrogen peroxide testing strip (Merckoquant 110011 or 110081) into the liquid for 1 second.
4. Wait for 15 seconds then read off the colour against the colour scale on the strip container. This scale goes from 0 to 25 mg H_2O_2 per litre. The colour will indicate a number.
5. Multiply the number by 5. The result gives the amount of H_2O_2 in micrograms (µg) as determined by the glucose oxidase from 1 g of honey in 1 hour at 20 °C. For example, a reading of 2 mg $H_2O_2 \times 5$ shows that 10µg of H_2O_2/g/hour at 20 °C are present.

Results

- If the number is greater than or equal to 10 mg per hour, this means that the HMF level will be lower than 40 mg/kg. Reliability: 95%.
- If the result is 0, this means that the honey has been heated too much or for too long.

Note:

Certain honeys may be difficult to analyse. That from thyme or mint contains high levels of vitamin C. The H_2O_2 oxidises this and so is reduced. The presence of katalase can also upset results. However, if you have a bad result you should send your honey to a lab for more accurate testing.

SECTION 35:
HARVESTING PROPOLIS AND POLLEN

Pollen harvesting

There is a large market for pollen, especially as a health food product, and some people have called it the perfect food. There are beekeepers in some countries such as Spain who dedicate their hives to pollen rather than honey collection and who make a living from it. Pollen is harvested by the use of pollen traps which are placed on the hive for this purpose.

Pollen traps

There are two basic types of pollen trap. The front-mounted trap attaches to the hive in front of the hive entrance. The bottom-mounted trap attaches underneath the hive. This is more effective and efficient. Both can be housed in a standard hive body.

Pollen traps should collect 50–80% of all the pollen entering the hive and can be obtained from any bee-supply store with attachment instructions.

Harvesting and post-harvest procedures

1. Choose hives brimming with bees, with good egg-laying queens.
2. Install the trap on the hive following the manufacturer's instructions. Begin to check the trap the next day for pollen.
3. Check the trap in the early morning before the bees leave the hive or in the evening after the bees have returned for

the night. Bring a large, plastic, food-grade container to deposit the pollen into.

4. Open the front entrance regularly. Worker bees need to clean the hive out, so open the trap at intervals – otherwise all their debris from the cleaning will drop into the trap!

5. Although it takes 2–3 days for the trap to fill, it should be emptied daily so that the perishable pollen can be processed.

6. Place the bee pollen from the food-grade plastic container into plastic freezer-storage bags. Promptly place the pollen in the freezer.

7. After a week, remove the pollen trap to allow the colony once more to collect pollen in the normal way.

8. Dry the pollen it contains to a moisture content of 2.5–6% (check with a pollen moisture meter). The recommended drying temperature is 45 °C, but merely heating the air is not sufficient in times of high humidity. It may also be necessary to dry the air before it is heated and forced through the pollen. A pollen drier is advised for small-scale production. These electrical devices can be purchased from bee-supply stores.

9. If you have no moisture meter, then try to break a pellet of pollen between your fingernails. If it won't disintegrate or is very difficult to crush, then the moisture content is about right.

10. Clean the pollen of debris. This is usually accomplished through a series of differently sized screens.

11. Pack and seal in airtight containers immediately after drying and cleaning.
12. Use within 12 months.

Notes:

- *Pollen should be frozen until you are able to dry it (to prevent mould and kill wax moth eggs as well as to preserve the pollen in a fresh state).*
- *Once dried, the pollen need not be refrigerated.*
- *Pollen should be kept frozen if it is not going to be dried.*
- *Each hive should produce ½–1 kg of pollen in the spring.*
- *Pollen traps are an excellent indicator of the hive's health. If production falls promptly, find out why.*

Harvesting propolis

Propolis is another side product of beekeeping and few, if any, commercial operators harvest propolis alone. With advances in our understanding of propolis and its uses in medicine, perhaps they should and maybe one day this moment will come. In the meantime, however, it is useful as a very saleable side product.

There are two basic methods to harvest this substance.

Scraping

The bees can propolise most parts of the hive, especially between frames and between boxes. Simply scrape off the propolis with a scraping tool and deposit in a container.

Disadvantages

However careful you are, this method usually includes small pieces of wood and paint in the harvest, and much time is taken removing these.

Using propolis screens

Essentially these screens rely on the fact that bees will propolise up any small holes found in the hive. Use them as follows:

1. Ensure the screen has holes in it around 3–4 mm in width.
2. Place the plastic flexible screen on top of the bars (where you would place a crown board if you use one) so that the bees will propolise the slots.
3. When the screen is full remove it and put it in a freezer.
4. When very cold or frozen, take it out of the freezer, flex the screen and out will pop the propolis. (In the commercial versions of the product, the slots in the screen have slightly sloping sides which aid the extraction of the propolis from the screen.)

Note:
Flexible plastic screens of all sorts can be purchased at DIY stores and garden centres. These can be ideal for use as propolis screens and are very much cheaper than the proper article.

Post-harvest

Remember to keep propolis in sealed containers. Propolis often contains wax and so it attracts wax moths (see Section 40).

SECTION 36:
WAX EXTRACTION

The production of beeswax is essentially a side product of beekeeping although one that can provide a useful, if sporadic, income. Beeswax is an ideal product because of the advantages of its collection, which are as follows:

- The processing of beeswax is easy. It can be simply moulded into blocks using any suitably sized containers.
- The transport and storage of beeswax are straightforward because no special packaging is required as it is not a food product.
- Beeswax does not deteriorate with age and so can be stored safely with no special precautions. Wax moths will have a go at it if they find it but will cause very little damage because there is little or no food value for them in pure wax.

Advantages of rendering wax

Rendering beeswax aids in apiary and hive management because it is an ideal way of recycling old black comb and replacing it with new foundation for the bees to pull out. Old comb should be replaced around every 3 years and you should institute a recycling period for this. For example, recycle one third of your old comb in the hive each season. Old comb can harbour problems and disease.

There are several ways of rendering down beeswax but the two most common for the hobbyist are solar wax extraction and steam wax extraction. Both methods offer advantages and disadvantages, as follows.

Solar wax extraction *(see Figure 6)*

Advantages

- These extractors can easily be made and kept in the field or taken into the field for the day.
- Plans exist on the web, and the device is simple to make and maintain.
- The power source is free.
- No need to keep attending to the extractor. It can be left to get on with it.
- They are cheap and easy to make and free to run.

Remember to keep the extractor at a slant towards the sun at all times for maximum benefit. Double-glaze the lid and paint the device black for maximum benefit.

Figure 6. Solar wax extractor

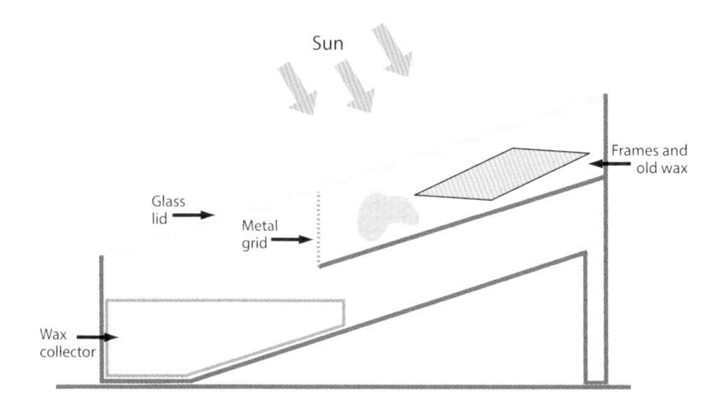

Disadvantages

- They won't extract all the wax from old comb.
- The wax container needs frequent emptying.
- They are not much use in the winter just when you have the time to render down all your old comb.

Steam/boiling-water extraction *(see Figure 7)*

Advantages

- The debris container is generally larger than in a solar extractor.
- The process extracts much more wax from old comb than solar extractors.

Figure 7. Steam/boiling-water extractor

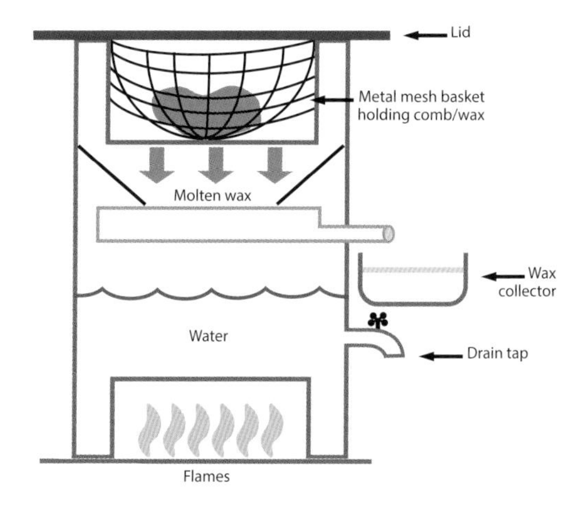

Disadvantages

- These extractors are more expensive.
- The power source adds extra costs.
- They are more dangerous due to boiling water and steam.
- They are not usually transportable into the field.

SF

Part G
Hive Checks

Introduction

– the need to plan

Planning is an essential part of beekeeping, and having a list of seasonal checks available in the apiary is a vital ingredient for success. If you know your own area, the climatic factors, local floral sources, periods of dearth and the strain of your bees, you can make up your own specific lists. But whether you use those supplied in this Field Guide or your own, always have one with you so that you can keep an eye on how things are progressing and so that you can prepare your seasonal plans well in advance.

Just one example will suffice to emphasise the need for careful planning. It is a surprise to most beekeepers when they expand their apiaries suddenly to realise just how many extra boxes they will need. One extra hive in the apiary needs one or two brood boxes and about three or four honey supers – a total of six extra boxes! Expanding your operation by just ten hives requires ten extra queen excluders, ten extra frame feeders and 60 extra boxes, plus 600 frames, 600 sheets of foundation or comb and extra storage room to store it all during the winter. Things multiply exponentially! This is where many beekeepers fail and their expansion plans are sunk by the enormity of it all – and the fact that they simply hadn't realised the extra costs and logistics involved.

Other questions arise. Where will you place the extra hives? Is there sufficient forage for extra hives in the area? Do you have enough time for it all? And so on. At least with a planner, a list of seasonal checks and a bit of common sense you can sit down and work out just what you can and can't do in the coming season (and what you can and can't afford). In this part two planners are supplied to give you a basis for working out your own according to your local environment and your own circumstances.

SECTION 37:

MONTHLY PLANNER (BASED ON THE NORTHERN HEMISPHERE)

January

- Ensure you have sufficient boxes and frames of comb or foundation to cover your plans for the coming new beekeeping year.
- Clean and repair equipment/hives.
- Check all stored comb for signs of wax moth.
- Place orders for queen bees now.
- Earmark equipment for spring queen rearing.
- Buy extra boxes, frames and foundation and other essential items for any expansion plans.

February

- If the temperature is around 13–16 °C and bees are flying, inspect the hives.
- Check that the bees have sufficient pollen and honey.
- Feed if you need to with invert sugar syrup, if available.
- Check pollen stores. If low, feed pollen patties.
- Remove mouse guards and replace with entrance reducers if still required.
- Ensure that all your extra items of equipment are available and ready for use.

March

- Watch for signs of swarming (see Section 18).
- Watch your hives to see if pollen is being carried in by foragers.
- Commence your queen-rearing plans as necessary (see Section 26).
- Prepare and place bait hives if needed (see Section 25).

April

- Start your queen-rearing activities this month (see Section 28).
- Commence swarm-prevention manipulations, such as reversing hive bodies and adding an extra brood box (if you have wintered on one only) (see Sections 18–25).
- Keep a good eye out for swarm preparations (see Section 18).
- Increase your inspections to around one a week/every ten days.
- Treat for Varroa.

May

- Start re-queening your hives (see Section 28).
- Super your hives if necessary.
- Keep up with swarm-prevention manipulations and maintain a close watch for swarming preparations.
- If necessary, carry out an artificial swarm (see Section 22).

June/July

- Keep placing on more honey supers if necessary.
- Extract spring honey if required.
- Look for signs of swarming and maintain your swarm-prevention measures.
- Order new queens for autumn re-queening.

August

- Harvest if required (see Sections 33 and 34).
- Start queen rearing for autumn queens (see Section 26).
- Move hives to winter sites if necessary (see Section 16).
- Autumn treatment for Varroa (see Section 49).
- Store all empty honey boxes, having treated the comb for wax moth (see Section 40).

September/October

- Carry out wintering checks (see Section 38).
- Check all the woodwork, lids, floors and frames.
- Reduce hives to two brood boxes only.
- Re-queen now if required (see Section 28).
- Feed if necessary (see Sections 57 and 58).
- Tie hives down if in windy areas or at least place a brick on the lid.
- Place mouse guards over the entrance.
- Ensure the hives can't be flooded.

November/December

- Watch hives, especially after bad weather conditions.
- Check for animal damage and flooding problems.
- Clean old comb off frames and render them down.
- Prepare new frames of foundation for the spring.
- Repair and clean old equipment.

SECTION 38:
WINTERING CHECKLIST

By checking the following aspects you will significantly increase the chances of a colony surviving the winter:

- The colony is headed by a young, productive queen.
- The colony is disease free (see Part H) and treated for Varroa (see Section 49).
- The colony has a sufficient reserve of bees. For cold climates, 10–15 frames.
- The colony has sufficient reserves of honey.

Northern climates

Average temperature <7 °C; minimum 40 kg in three brood boxes. Usage and research have found that, as a rule of thumb, the following amounts are required: three boxes (10+ kg in bottom box, 15 kg in middle and rest in top box). Average temperature (UK) –4 to +10 °C; 15–30 kg.

Southern climates

Average temperature >8 °C; 15 kg; 15–20,000 bees. These reserves should be built up with sugar syrup (preferably invert) if insufficient honey has been left.

Ensure the following:

- The honey reserves are properly organised. In cold areas, many authorities recommend a three-box wintering unit with reserves in all three, but with none of the boxes being honey bound. In mild areas where inspections can be carried out, ensure there are always at least four combs of honey available as well as pollen.

- The colony has sufficient pollen reserves (something many hobby beekeepers ignore or fail to understand). Again, this pollen must be available in or next to the cluster. The provision of pollen substitute (see Section 57) near to the cluster 4–5 weeks before the availability of natural pollen will stimulate brood rearing. A lack of pollen will cause colony dwindling in late winter and early spring.

- The colony is securely housed and protected from predators (mice, woodpeckers, cattle, etc.) and extremes of wind/snow drift, etc., by the provision of wind breaks.

- The provision of upper ventilation to prevent condensation and for escape if the lower entrance is blocked.

Notes:

- *Bees don't die of cold. They starve.*

- *Bees don't try to heat the hive – they maintain the cluster temperatures.*

- *Autumn feeding of sugar syrup can be used to administer Fumigillin in those areas where it is permitted.*

Part H
Pests and diseases

Introduction

– a warning and reminder

The field diagnosis of diseases can be difficult. Some, such as chalk brood (see Section 43), are more readily identified than others, such as European foul brood (EFB – see Section 42), the visual symptoms of which can be confused with those of other problems. In many cases (e.g. with foul brood) it is easier to identify diseased colonies if you know what a healthy one looks like.

If you have doubts concerning a field diagnosis, seek advice from someone competent or contact the statutory authority. It is inadvisable just to do nothing in the hope that it will go away. This is especially important in cases of American foul brood (AFB) and EFB, which in some countries/states are not notifiable diseases.

This part of the guide starts with a disease field-identification table that gives a brief description of the problems and that directs the reader to each relevant section of this part of the book for more details. Wax moth begins the list because the presence of wax-moth damage in the hive is usually a sign that another problem/disease is present. In fact, for many beekeepers wax-moth damage is often the first sign they see of a problem with their colony.

Integrated Pest Management (IPM)

When dealing with the diseases of any livestock it is usually advisable to use an IPM strategy. This is the management of pests employing a combination of methods that include economic, ecological and toxological factors while emphasising biological controls and economic thresholds.

The basic components of an IPM programme include:

- prevention and awareness;
- observation and monitoring; and
- when necessary, intervention.

Your apicultural extension office will be able to advise you on an IPM strategy which will better help you control problems.

SECTION 39:
DISEASE FIELD IDENTIFICATION

Poisoning (see Section 17) has also been included in the table because it can resemble other diseases and problems.

Table 14 Disease field identification

Disease	Identification
Wax moth (see Section 40)	White lines running through the combs Dark webbing and faecal matter present Combs often stuck together by webs Larvae crawling through the combs Pupae sticking to the combs and all crevices In bad cases combs fall apart Emerging brood dead (galleriasis)
AFB (see Section 41)	Combs may show a pepperpot appearance Honey may be stored in the brood nest among the diseased cells Cappings dark brown and sunken Cappings often perforated Decaying larvae will rope at sunken capping stage (push a match through the cap) Noticeable foul/different smell Dry scale shows pupal tongue projecting to the centre of the cell
EFB (see Section 42)	Larvae are an off-white colour The trachea are easily visible The larvae adopt unnatural positions The larvae appear 'melted down' The comb may have a pepperbox appearance. Honey may be stored in the brood nest. Larvae may 'rope', but not as far as AFB Cappings may become sunken, perforated and discoloured Noticeable foul smell

Chalk brood (see Section 43)	Open cells with hard chalky-white remains Infected cells scattered over comb Infected larvae covered in fungus Chalk-brood mummies on alighting board Mummies can be tapped out of cells
Sac brood (see Section 44)	Cell cappings may be perforated Open cells show darkened larval head Open cells show bloated abdomen Larvae can be withdrawn with a pin Scale resembles a Chinese slipper or gondola
Chilled brood (see Section 45)	Brood of all ages dies at the same time Larvae go dark (almost black) in colour Brood often affected most at periphery of the brood nest
Bald brood (see Section 45)	Cappings over healthy brood removed Small rim often built around the cell
Nosema (apis and ceranae) (see Section 46)	Colony fails to build up and may reduce in size, despite favourable conditions Field test available *Apis ceranae* 'may' be closely associated with colony collapse disorder (see Section 53) and can kill a colony very rapidly
Dysentery (see Section 47)	Faecal spotting around the hive entrance Severe cases show this all over the hive Defecation within the hive, especially after a long winter
Virus diseases (see Section 48)	Many moribund bees on/around the hive; bees don't react if you prod them Affected bees unable to fly Bees noticeably blacker Bees have a hairless, greasy look Distended abdomens Pile of dead bees in front of the hive Bees on top of the pile still moving (can be many, but usually fewer than those caused by insecticide spray poisoning) Affected bees refused entry to the hive

Varroa destructor (see Section 49)	This mite is associated with virus diseases and unless treated will lead to colony collapse. In severely infested colonies, symptoms include a rapid reduction of adult bees; some deformed bees, especially wings; and brood disease-type symptoms. The mite is small but can most easily be seen on uncapped drone brood. It is red/brown in colour and 1.5 mm wide and 1.1 mm long
Tropilaelaps clarae (see Section 50)	Similar to Varroa but smaller, elongated and not crab shaped. Can cause rapid colony loss
Tracheal mite (see Section 51)	Commonly known as Acarine in the UK. Few if any symptoms visible in the field. Symptoms thought to be associated with this mite may be those of virus disease
Parasitic mite syndrome (PMS) (see Section 52)	Symptoms difficult to interpret but can include those similar to foul brood (see Sections 41 and 42) but not ropiness or bad odour Cappings often perforated or uncapped like AFB Affected brood can be anywhere in the comb and of any age Colour (dull white/yellow and later with brownish spots) is the best way to distinguish from foul brood PMS is always associated with Varroa (see Section 49)
Poisoning (see Section 17)	Pile of dead bees at the hive entrance (all bees usually dead) The quantity of dead bees is usually larger than the number caused by disease Returning field bees denied entry Much reduced hive population Usually affects all hives in the apiary Some pesticides cause slightly affected bees to become very aggressive Others cause bees at hive entrance to tumble about, spin around, tremble, unable to walk (fly-spray effect)

Poisoning (later effects) (see Section 17)	Spotty brood pattern due to insufficient adult bees to look after the brood Dead larvae removed Chilled brood (see Section 45) similarly caused and usually at the edge of a brood area Colony stress leading to EFB (see Section 42), chalk brood (see Section 43) and sac brood (see Section 44), the symptoms of which may occur up to 6 weeks after the event
Colony collapse disorder (see Section 53)	The hive is abandoned by the workers, often leaving brood and stores. This can happen very quickly and has caused massive losses in the bee industry
Small hive beetle (see Section 54)	A relative newcomer to *Apis mellifera* colonies. The wax and comb are infested and the beetle and its larvae eat through both wax and brood destroying the nest and often causing the hive to be abandoned

SECTION 40:
WAX MOTH

Where there is beeswax, there will be wax moths, and these can be effectively protected only by healthy colonies that can control the pest or by other protective measures in stored comb. It is not uncommon to see one or two moths in the more remote corners of a healthy hive, but any signs of actual wax-moth damage mean that there is a problem in the colony and so you should look for the underlying cause(s).

Identification

The greater wax moth, or *Galleria mellonella* (1.3–1.9 cm long), is usually around in most beehives and the bees usually repair any damage as soon as it appears. It is for this reason that most beekeepers are unaware of it until the bees lose their ability to defend themselves and the moth larvae take over. The problem is more acute in warmer countries because the conditions favour continuous reproduction.

16. *Greater wax moth*

The lesser wax moth *(Achroia grisella)* is smaller and more silvery in appearance, and its larvae are correspondingly smaller than those of the greater wax moth.

Symptoms in order of severity:

- White lines under the comb surface, running through the brood cappings. Exposed brood shows no sign of disease (low infestation).
- Webbing and faecal matter disfiguring the combs. Webbing often dark in colour.
- The amount of webbing means that the combs are stuck together. Much faecal matter.
- Boat-shaped depressions in the woodwork of frames and the hive caused by the larvae pupating. Some may even bore right through frames, which can resemble Swiss cheese.
- Comb falls apart, resembling sawdust (see Photograph 17)
- Emerging brood unable to emerge fully from the cell and so they die *in situ* (Galleriasis) (Flottum and Morse, 1997).[1]

Treatment

- If only slight damage, clean up the combs and look for the reason why the colony succumbed. Treat accordingly.
- If severe, strip all the frames, clean them up and scorch them with a blow torch. They can then be used again. Burn the debris, then scrape the hive body, floor, lid and supers and blow torch them. You will find pupae in many of the cracks and crevices in the hive.

17. *Wax-moth damage: comb falls apart, resembling sawdust*

Prevention and control

- The spray application of *Bacillus thuringiensis* (available from bee-supply firms) will prevent damage, or further damage. This can be applied before storage or before putting comb into the hive. It does not affect bees or beekeepers, contaminate the honey or wax, or cause any other damage.[2]
- In warm climates or during warm winters, keep the number of supers on the hive to the minimum (see Pointers below).
- The sulphur fumigation of stored combs is effective.

Pointers

- Visual symptoms of damage readily identified.
- A few larvae are often seen in a hive. This is normal.
- Larvae may be of two sizes (greater/lesser wax moth).

- Wax moth prefer used comb.
- From the outside, everything may appear normal.
- The presence of wax-moth damage and larvae in large numbers usually indicates a problem with the colony.

Notes:

- *Colonies usually succumb to wax moth because of disease or queen failure (i.e. the colony is declining).*
- *If you find wax-moth damage, look for the problem.*
- *If the colony has been destroyed and no sign of disease or other problems remain, then it is best to sterilise all hive parts, clothing and equipment before reuse. A blow torch is useful for this.*
- *Stored comb that is unprotected is especially at risk.*
- *In warm climates or warm winters, colonies may succumb to wax moth if, for example, an unused super is left on during a period of colony decline (winter). With fewer bees available the colony will be unable to control the onslaught, even though the colony is in other ways healthy.*

SECTION 41:
AMERICAN FOUL BROOD (AFB)

AFB is the most serious of the brood diseases. It is a highly infectious bacterial disease and can be spread by drifting bees, robbing and by the beekeeper moving from an infected hive to other hives during inspections. Colonies that have AFB must be destroyed. The bees must be killed and brood frames must be burnt. Woodwork other than the frames may be saved according to state or national law but usually must be sterilised effectively. In many countries outbreaks of AFB must be reported to the appropriate authority and it is they who deal with the problem.

Cause

AFB is caused by the spore-forming *Paenibacillus* larvae (formerly classified as *Bacillus* larvae).

Identification

- Combs may or may not show a 'pepperpot' appearance.
- If they do, honey and pollen may be present within the brood area.
- The cappings become discoloured (dark brown) and sunken.
- The cappings are often perforated (see Photograph 18).
- The larva within the cell becomes discoloured.
- At the sunken cappings stage, the brown remains of the larva may be drawn out like a thread for about 1–2 cm (ropiness test) (see Photograph 19). Prod through the capping with a match.
- Later, the larvae become darker and tacky.

- When the larva dries out, the larval scale lying along the lower side of the cell may show the pupal tongue projecting from the scale to the centre of the cell. This is a very distinctive feature.
- Colonies with much AFB smell foul. This really is noticeably different from that of a healthy colony.

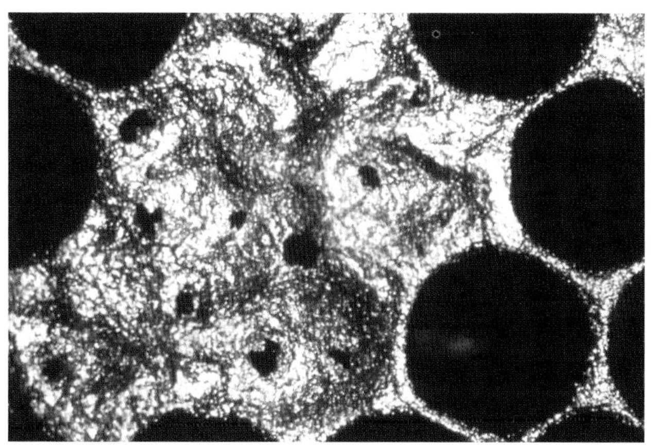

18. *AFB: cappings perforated*

Treatment, prevention and control

Treatment by beekeepers is often limited or prohibited by national or state regulation. If in doubt, contact the statutory authority for beekeeping matters.

- Prevention and control can be effected using oxytetracycline, which comes under various trade names. *The use of oxytetracycline is inadvisable without specialist advice, both in the quantities given and in the method by which it is given. It can hide and not cure the problem, and extended use can lead to bacterial resistance* (Delaplane, 1998a).

- Burning of the hives, including the bees, is very effective.
- Burning of the frames and bees and the scraping and blow torching of the hive parts is also effective.
- Irradiation of hive parts and frames sterilises them (Hornitzky, 1994; Francis et al., 1995).

19. *AFB: the ropiness test. To the right is a clear example of the AFB scale showing the protruding pupal tongue*

Pointers

- This is probably the most serious brood disease.
- It is highly infectious.
- Its name bears no relation to its geographical spread.
- The early stages are not easy to identify.
- The larvae die after the cell has been capped.

- It appears that certain methods of providing antibiotics to the bees may be leading to resistance to oxytetracycline, and inexperienced use may contaminate honey (Hornitzky, 1994).

Notes:

- *If you leave AFB unchecked, it will destroy the colony and spread to others due to factors such as drifting, robbing (see Section 13) and contamination of the beekeeper's equipment.*
- *If you have left AFB unchecked then, before you go near another hive, sterilise all your equipment (hive tool, etc.) with a blow torch and wash all your clothing. Buy new gloves and burn the old ones.*

SECTION 42:

EUROPEAN FOUL BROOD (EFB)

Like AFB, EFB is a bacterial disease. The causal agent of EFB is the bacterium *Melissococcus pluton*, which infests the guts of bee larvae. Although considered less damaging to a colony than AFB, it should never be underestimated and should be attended to if and when detected. The bacterium does not form spores though it can over-winter on comb. Because it doesn't form spores it is not as infectious as AFB and, if it is caught in its early stages, the colony can usually be saved.

It is often considered a 'stress' disease – a disease that is dangerous only if the colony is already under stress for other reasons, such as frequent moves, other disease problems, pesticide poisoning and so on but, if the colony is given a chance to build up, it can usually survive.

Identification *(see Photographs 20 and 21)*

- The larvae change from white to a creamy off-white colour.
- Because of this, the tracheae become easily visible.
- The larvae adopt unnatural positions in the cell and are not neatly coiled.
- The larvae appear to lose definition and appear melted.
- The presence of normally capped cells may lead to a pepperpot appearance of the comb.
- In the later stages the decomposing larva may 'rope'.
- The cappings of cells of larvae that are sealed over before death become discoloured, sunken and perforated.

- The later stages of the disease produce a 'foul' smell.
- The outbreaks usually become evident in the spring/early summer when the colony is growing rapidly.

20. *EFB: appearance of the brood comb*

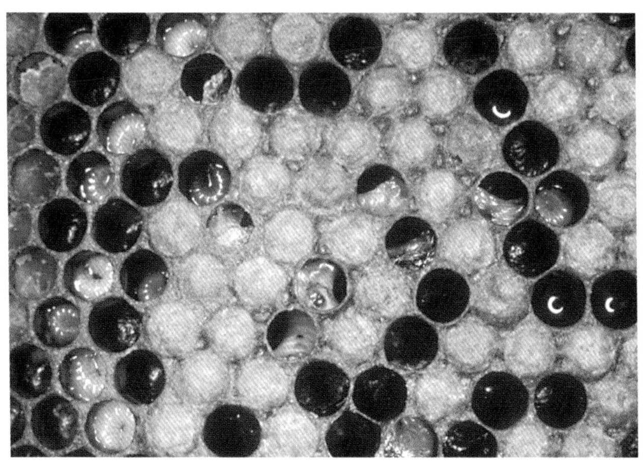

21. *EFB: close-up of diseased larvae – some discoloured, some with 'melted down' appearance, next to healthy brood*

Treatment, prevention and control

Treatment of EFB by beekeepers is often limited or prohibited by national or state regulation. If in doubt, contact the statutory authority.

- Treatment of EFB is not usually necessary if the infection is light and the colony is strong (but you may still have to report it).
- Prevention and control can be effected using oxytetracycline, which comes under various trade names. This should be applied early in the season (i.e. usually before EFB is noticed). As noted above, the use of oxytetracycline is inadvisable without specialist advice and may need permission.
- Burning the hives and bees is effective.
- Burning the frames and bees and cleaning and blow torching the hive parts is also effective (especially frames containing dead brood).
- Re-queening, with its associated break in the brood cycle, can be effective, but some research shows that the brood-less period may be of little value and may even be counterproductive (Bailey and Ball, 1991).
- Clean, empty comb can be fumigated with acetic acid.

Pointers

- The larvae die of starvation.
- The larvae usually die before being sealed over.
- Larvae can die after being sealed over.

- EFB's name bears no relation to its geographical spread.
- Identification is difficult in colonies used for queen rearing because they are generally well or overfed. The larvae are thus able to obtain enough food to overcome the parasitic nature of the bacteria.

Note:
EFB can exactly resemble half-moon syndrome (HMS), which is named after the half-moon shape of the larvae at one stage of the syndrome. HMS does not exhibit the bacteria of EFB. HMS may be a queen problem. What have you got? EFB or HMS? It has been suggested that many cases of EFB, especially those that clear up swiftly and have little effect on the colony, are, in fact, HMS and that the beekeeper simply didn't realise it. Bacterial analysis is usually the only way of telling the difference at this stage.

SECTION 43:

CHALK BROOD

Chalk brood is a fungal disease caused by the fungus, *Ascosphaera apis*, and it affects unsealed and sealed brood. It is fairly easy to recognise especially in its later stages due to the floor and front of the hive being littered with small hexagonal blocks of chalk-like material.

Identification *(see Photographs 22 and 23)*

- Open cells filled with hard, chalky-white remains, often hexagonal in shape.
- Later, these 'mummies' shrink and become harder.
- Grey discoloration to the rear end of the 'mummy' if removed from the cell.
- Larvae may be covered in a grey/white mould.
- Infected cells are usually scattered throughout the comb.
- Chalk-brood 'mummies' on the alighting broad and/or the bottom board of the hive.
- Tap the frame lightly downwards against a solid surface and many of the hard mummies will fall out (this is a fairly conclusive diagnosis).

Treatment, prevention and control

There is no real satisfactory treatment for this disease. The following may help to reduce its occurrence:

- Avoid opening hives in cold weather.

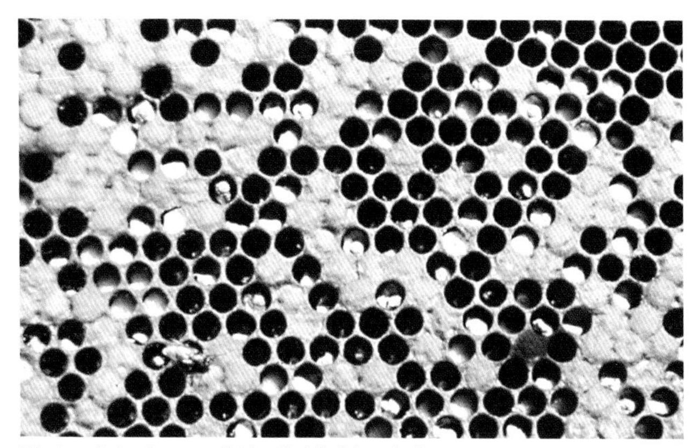

22. *A comb showing an easily identified chalk-brood infection*

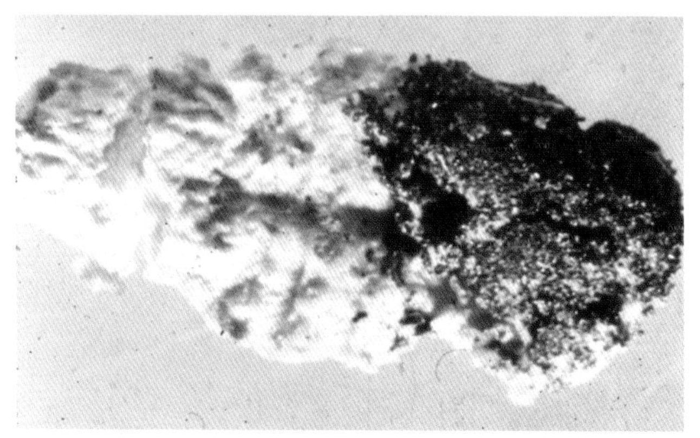

23. *A chalk-brood mummy*

- Avoid manipulations that give weak colonies too much brood to rear during the early part of the year (brood chilling).

- Avoid carrying out swarm-control manipulations too early in the year (brood chilling).

- Avoid manipulations that can cause stress to the colony.
- The addition of bees to a weak hive may help the situation.
- If the situation is serious, re-queening from a 'clean' hive may work.

Pointers

- The larvae die after they have been capped.
- Chalk-brood spores are often noted in healthy colonies.
- Adult bees usually remove the cappings of affected larvae in order to get rid of the remains.
- Chalk brood reaches its peak in the early part of the season.
- It is not usually a serious disease but is believed to be stress related.

Notes:

- *Chalk brood can easily be spread by robbing, drifting and the beekeeper's manipulations.*
- *The spores of* Ascosphaera apis *are ingested with the brood food as the larva feeds itself. On the germination of the spores, the fungus breaks out of the larva and covers it with a white mycelium.*
- *The spores remain viable for years. Consequently, the infection source could be present in the cells used to rear brood.*

Section 44:
SAC BROOD

Sac brood is a viral disease *(Morator aetatulae)* that does not usually cause severe losses. It occurs mainly early in the brood-rearing season when the ratio of brood to bees is high. Most beekeepers don't notice it because it mainly affects only a small percentage of larvae. Adult bees detect and remove infected larvae very quickly and so, if the beekeeper does notice the problem, it is usually because it has progressed to a stage where the workers can no longer control it. Therefore, by the time the beekeeper observes the symptoms, the disease may be too severe for the adult worker population to handle.

Part H

Identification *(see Photographs 24 and 25)*

- Cell cappings may be perforated.
- Cells may be open, showing a darkened larval head, but if removed from the cell the abdomen is bloated, resembling a watery sack, which gives the disease its name. This can be pulled out with a pin for certain identification.
- The larva eventually dries to a scale, resembling the shape of a Chinese slipper/gondola. This is very distinctive and can be easily removed by the bees (unlike the AFB scale; see Section 41).
- After a few weeks, the larval remains are not infective.

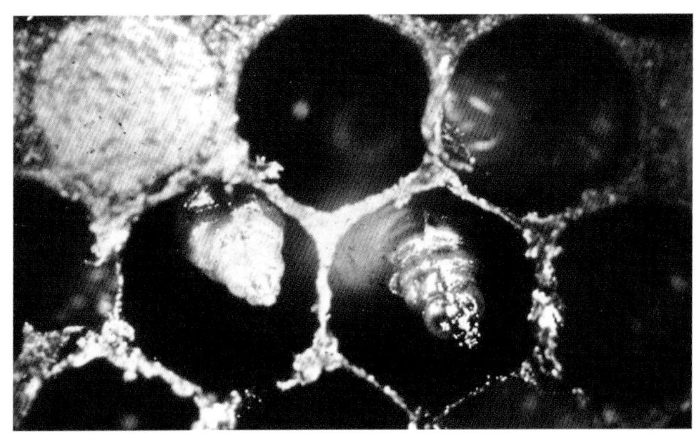

24. *Sac brood: close-up showing appearance of diseased larvae in the cells (pointing upwards)*

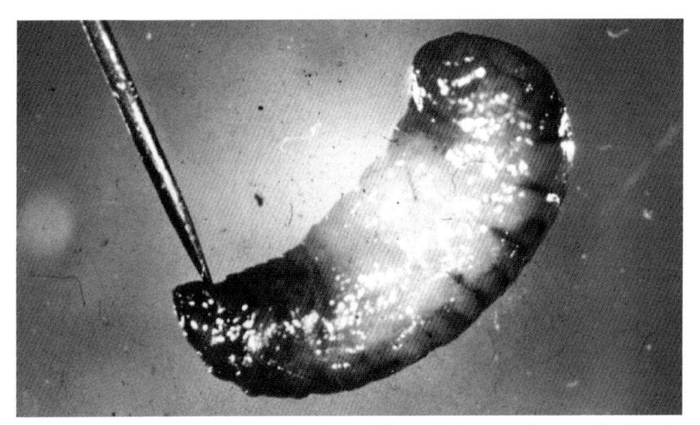

25. *Sac brood: bloated appearance of the diseased larva, which are easily removed with a pin*

Treatment

- There is no treatment but, if the disease is serious, re-queening from a 'clean' colony can work (Allen and Ball, 1996).

- Colonies suffering from the virus usually recover spontaneously when a good honey flow starts. At this time, there are fewer infected adults to pass on the virus to the brood.

Pointers

- The larvae die after the cell is capped (failure to pupate).
- Adult bees ingest the virus when cell cleaning. It then multiplies in them, but without causing apparent illness.
- Over-wintering adult bees which are carrying the virus may cause a new outbreak of the disease in the following spring.

Notes

- *In the UK, it is believed that most colonies carry some infection.*
- *It may cause more harm to adult bees than is apparent.*
- *The virus is probably fed to the young larva by the nurse bees in the brood food.*
- *It multiplies rapidly within the larva until it causes death. Then the house bees cleaning out the cells probably distribute the virus to other larvae within the hive.*

SECTION 45:
CHILLED BROOD AND BALD BROOD

Chilled brood

Identification

- This is recognised easily because brood of all ages die at the same time.
- The brood most affected is usually at the periphery of the brood nest.
- Larvae turn black before drying up.

Causes

- Too few bees available to keep the brood warm (e.g. after pesticide poisoning) (see Section 17).
- Nucs kept in the same apiary as they were made may lose too many flying bees to their original hives.
- Other manipulations may cause a loss of bees or isolation of the brood (e.g. spreading brood) (see Section 31).

Treatment

Simply ensure that colonies are always well populated and take proper care of nucs (see Section 21).

Bald brood

Identification

- Identified when some of the cappings of normal healthy brood are removed without harming the larvae.
- A small wall is often built around the cell.

Causes

- Can be caused by the greater wax-moth larvae (see Section 40) chewing through the cappings. The bees remove the rest of the capping to clear away the silk. The damage is usually in straight lines.

- It may also be an inherited trait among a few bees in the colony.

Pointer

The larvae usually pupate and emerge normally.

Treatment

None is usually necessary.

Note:
Occasionally, some of the symptoms of these brood diseases can be confused with some of those of AFB (see Section 41). If in doubt, seek advice.

SECTION 46:
NOSEMA

Nosema apis

Nosema apis is a unicellular parasite of the class of microsporidian *(Microsporidia)* which are now considered to be fungi. *N. apis* has a resistant spore that withstands temperature extremes and dehydration. It is a very widespread disease of honeybees and, when spores are eaten by adult bees, they germinate and invade the gut wall. Here they multiply and produce more spores which are passed out in the waste.

Nosema is common in the spring and autumn. Many beekeepers treat the condition with a substance called fumagillin, the trade name of which is Fumadil-B, added to an autumn feed of sugar syrup. Treatment with this antibiotic (prepared from *Aspergillis fumigatus*, the causative agent of stone brood!) inhibits the spores reproducing in the ventriculus but it does not kill them. The effect is that the colony is allowed to recover. Fumagillin is not permitted in some countries.

Identification

- There are no specific, external visible symptoms *(Vida Apicola,* 1998).
- A failure to build up in spring in ideal conditions could indicate Nosema.
- There is now a field test available which can give an idea of the presence of Nosema but which is not definitive (see Photographs 26, 27, 28 and 29).

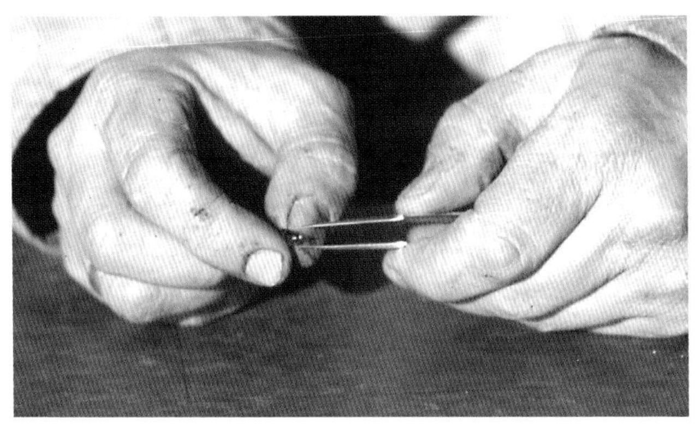

26. *Nosema field test: grasp the sting and last segment with the tweezers*

Treatment, prevention and control

- Good management practices are a sound way of reducing the incidence of Nosema.
- Ensure that the colony goes into winter with a young, prolific queen and many young bees.
- Ensure that the honey and pollen stores of colonies are adequate.
- Feed fumagillin. If fed in sugar syrup to over-wintering colonies, this can reduce markedly the incidence of Nosema during the following spring.

Pointers

- Nosema is a disease of adult honeybees.
- It is very widespread.
- It is believed to be a cause of queen supersedure.
- It is less common in warm climates with mild winters.

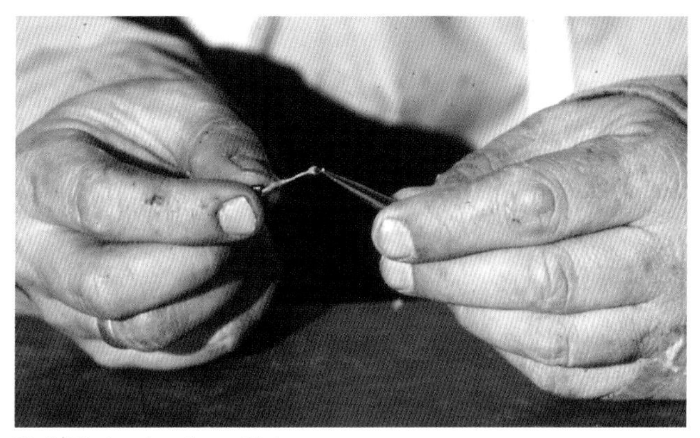

27. *Pull firmly and gently out of the bee*

28. *Lay out on a sheet of white paper*

Notes:

- *Crawling bees around the entrance to a hive may indicate other problems, such as virus disease (see Section 48). This is not a symptom of Nosema.*
- *If feeding medicated syrup, do not follow this with non-medicated syrup. This will dilute the effect and markedly reduce the beneficial activity of the chemical.*

29. *A healthy specimen: the mid-gut should be tan coloured and wrinkled*

- *Dysentery (see Section 47) is a major cause of spreading Nosema disease but is not in itself a symptom of Nosema.*

Nosema ceranae

This variant of Nosema came to the attention of European and American beekeepers in the late 1990s but it wasn't until 2004, when massive colony losses in Spain prompted scientific investigation into the phenomenon, that most beekeepers heard about it. Like *Nosema apis*, there are no specific external physical symptoms in the field except a rapid decline in the colony strength, which can lead to colony collapse very rapidly.

Pointers

- A major difference between the two types is the speed at which *N. ceranae* can cause a colony to die.
- Bees can die within 8 days after exposure to *N. ceranae*, which is much faster than bees exposed to *N. apis* alone.

- Foragers leave the colony and are too weak to return. They die in the field, leaving behind a small cluster and a weak colony – very similar to the symptoms of CCD.

- There is little advice on treatment but it has been suggested that the most effective control of *N. ceranae* is the antibiotic fumagillin recommended for *N. apis*.

- *N. ceranae* was first described in 1996 and was identified in 2004 as a disease of *Apis mellifera* in Spain.[3]

(See also CCD, Section 53.)

This way of diagnosing Nosema in the field (which is easier than it looks) is a good test for the presence of Nosema in your bees, but it is not an absolute and there can be some contradictory signs. The best tests are carried out in the laboratory and so, if you have doubts, collect up 30–40 bees, place them in the fridge for a few hours and then send them off to your local bee lab.

You will need the following items:

- A good pair of fine, pointed tweezers.

- About 30 bees from each colony checked.

- A sheet of white paper.

Procedure

1. Remove the bee's head. This severs the mid-gut from the head. (The bee may continue to struggle.)

2. Grasp the very last segment and sting with the tweezers and, gently holding the thorax with your other hand,

slowly but firmly pull the sting and last segment away from the bee (see Photograph 26).

3. The rectum and mid-gut will follow (Photograph 27). Keep pulling slowly and firmly. Don't pull suddenly or too hard, otherwise something will break and you'll need to start again (the bee may still be struggling).

4. After the mid-gut has emerged, hold it over a white sheet of paper (see Photograph 28). The mid-gut is easily seen.

5. Study the mid-gut. If it is tan coloured and wrinkly, it is healthy (see Photograph 29). If white and smooth, it probably has Nosema.

Notes

- *The bee will often struggle until the mid-gut is finally pulled from it. This can be alarming.*
- *If the bee stings you prior to the operation (which the bee in the photographs did), the method will still work.*

SECTION 47:
DYSENTERY

Dysentery is not a disease but a symptom of a disease or nutritional disorder.

Identification *(see Photographs 30 and 31)*

- Faecal spotting and soiling on and around the hive. If severe, the hive may be covered with it.
- Defecation within the hive, especially after a prolonged winter during which the bees are unable to take cleansing flights.

Treatment, prevention and control

As dysentery is a symptom, look for the cause of the problem. It is most likely to be contamination of the food supply or unsuitable winter stores. Ensure that any feed given before the winter is not contaminated.

Pointers

- Dysentery is caused by excessive water accumulation in the rectum.
- Dysentery can spread Nosema (see Section 46) but is not necessarily an indication of Nosema.

Note:
The prevention of dysentery, like so many other problems, is basically down to sound beekeeping, such as good feeding practices.

30. *Dysentery: a mild case after a prolonged winter*

31. *A severe case of dysentery can look like this*

SECTION 48:

VIRUS DISEASES

This is a very complex subject. Until the 1980s the specific viral infections of honeybees were generally thought to be of little threat to colonies.

The term 'virus diseases' covers a great many viral problems, including Kashmir bee virus and Israeli acute paralysis virus, which seem to have been given a new lease of life by the introduction of Varroa destructor that is believed to be a viral-vectoring agent. Varroa has also been seen as an activator of virus multiplication in infected individuals. Although there is no prescribed set of symptoms confirming virus diseases in the field (diagnosis being carried out in the lab using polymerase chain reaction (PCR) analysis), beekeepers may observe symptoms that may identify viral infection of one sort or another (see also PMS, Section 52).

Identification

These signs can be seen on many occasions but they can also be confused with other problems:

- A general weakening of the colony with no apparent brood disease or mite problems (could be Nosema).
- Bees crawling around the alighting board and the hive generally in a moribund condition, sometimes in thousands (could indicate starvation).
- These bees sometimes group together.
- They will not react if you prod them.

- They are usually unable to fly.
- They will often appear 'blacker' in colour than other bees. This is very noticeable (black bees only) (*Vida Apicola*, 1998).
- Older bees are often hairless and have a greasy appearance.
- There will often be a pile of dead bees on the ground in front of the hive.

Treatment

Various methods can be used to fight these diseases but they are not generally applicable as field treatments. These methods include:

- the selection of Varroa-tolerant bees;
- the prevention of new pathogen introduction; and
- RNA (genetic engineering) interference.

Generally, at a beekeeper level, you should do the following:

- Keep Varroa levels to a minimum with a good treatment regime (see Section 49).
- Reduce stress levels (e.g. less moving/handling).
- Re-queen because circumstantial evidence points to a genetic component.
- Reduce the effects and incidence by using good, hygienic hive/apiary management practices.
- Send a sample of affected bees to your national or state bee laboratory for definitive analysis.

Pointers

- Many bee viruses exist in hives that cause few problems, but when associated with heavy infestations of Varroa they can cause significant damage. Control Varroa and you can better control virus damage.

- Remember that some of the symptoms of virus disease can indicate starvation or poisoning (see Section 17). For starvation symptoms, look in the hive. For poisoning, look at the pile of bees. Usually they will all be dead and equally decomposed.

(See also Johansson and Johansson, 1978; Allen and Ball, 1996.)

Viruses affecting honeybees

Some of the viruses you will hear about in beekeeping meetings and read about in scientific research papers on diseases are listed below. They are included because they do produce a series of symptoms noticeable in the field and because they are also associated with more noticeable problems, such as PMS (see Section 52) and CCD (see Section 53). They are as follows:

- **Acute bee paralysis virus (ABPV or APV):** this is considered to be a common infective agent of bees. It belongs to the same family as the Israel acute paralysis virus, Kashmir bee virus and the black queen cell virus. It is frequently detected in apparently healthy colonies. This virus may play a role in cases of the sudden collapse of honeybee colonies infested with Varroa.

- **Israel acute paralysis virus (IAPV):** a related virus described in 2004 that is thought to be associated with CCD.

- **Kashmir bee virus (KBV):** related to the viruses mentioned above. KBV is currently positively identifiable only by a laboratory test. Little is known about it yet.
- **Black queen cell virus (BQCV):** this is another member of the same virus family. As its name implies, BQCV causes the queen larva to turn black and die. It is thought to be associated with Nosema.
- **Deformed wing virus (DWV):** this is thought to be present in hives. On its own, DWV tends to remain at low levels in the bees and exists as a symptom-less, low-grade infection. However, when the bees are under stress and the virus concentrations rise, bees emerge from the pupal stage with a variety of deformities in proportion to the quantity of virus in their systems, such as damaged appendages; deformed, ragged wings; shortened, rounded abdomens; discolouring; and paralysis. It is typically associated with Varroa destructor and it has been suggested as a contributing factor to CCD, along with a variety of other causative agents. In fact, DWV in Varroa-infested colonies is now seen as one of the key players in final colony collapse.

Two other commoner viruses of the 18 or more that can affect honeybees are:

- chronic paralysis virus (CPV); and
- cloudy wing virus (CWV).

SECTION 49:
VARROA DESTRUCTOR

(See also Varroa mite number calculations, Section 60.) It is Varroa destructor (previously misnamed as *Varroa jacobsonii*) that has caused so much trouble for many of the world's beekeepers since it jumped from its natural host, the far-eastern honeybee *Apis cerana*, to the western honeybee, *A. mellifera* – a bee that didn't know what to do about it.

The subject of Varroa and its effects on honeybees is vast and one that we are only just beginning to understand. Its relationship with other 'syndromes', such as PMS and CCD, is the subject of furious research on a mite that has had such an impact on the economy of beekeeping.

Identification and assessment of infestation

(see Photographs 32, 33 and 34)

- For the visual identification of mites on adult bees, the mite is a red/brown colour and 1.5 mm wide and 1.1 mm long. It can easily be missed.
- Mites can more easily be seen by uncapping drone brood (advanced pink-eye stage) (see the 'Rules of thumb' section below).
- An easy test of 3–500 bees can be carried out in the field.
- Natural daily mite fall can be used to assess Varroa infestation (see 'Rules of thumb' section below).
- A sticky Varroa floor and Varroa control strip can be used in assessing mite infestation.

- In severely infested colonies, a rapid reduction of adult bees occurs.
- In severely infested colonies, some or many adult bees have deformed, ragged wings and deformed abdomens (see Photograph 34).
- In severely infested colonies, foul brood-type symptoms may be seen. Be careful to distinguish between them (see also Sections 41 and 42).

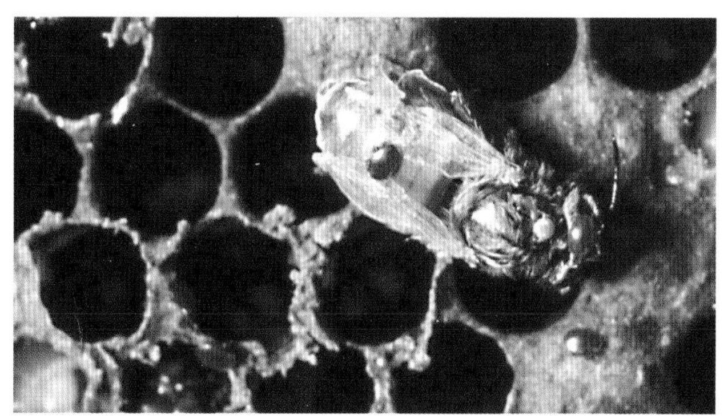

32. *Varroa on an adult bee*

Part H

Treatment, prevention and control

Treatment reduces the infestation to a level the colony can tolerate. A summary of the treatments and controls available is given below. If in doubt, use an authorised miticide.

The following treatment regime is a basic suggestion. Each beekeeper will develop their own regime according to their experience and circumstances.

33. *Varroa on larvae*

Authorised miticides

Various treatments designed for use in beehives are available (see below). These are easy to use and come with full instructions. Check with your local advisory service which ones are legal in your area. Ensure that any you use are authorised in your state/country (beware of mite resistance to some of these; an easy mite-resistance test is given below).

Timing and suggested treatment

This is a complex area of research. There are several prevention and control strategies available to beekeepers and you should always keep up to date with the latest innovations. Similarly, always ask for guidance from your local adviser.

The timing of the treatment is important. Good treatment in spring (when large amounts of brood are providing ideal Varroa conditions) is important. A lack of control now will lead to colony collapse in late summer. Use an authorised miticide, such as Apistan or Bayvarol.

34. *Typically deformed wings on a young adult bee*

To protect against Varroa invasion, treat in early autumn (after the harvest). Use an organic product, such as oxalic or formic acid. This will help to slow down mite resistance (Delaplane, 1998b).

Rules of thumb in the field

Drone-brood inspection – 100 cells examined and tested:
5% with Varroa = low infestation
25% with Varroa = high infestation

Natural mite-fall inspection

Daily fall:
May – 6 mites
June – 10 mites
July – 16 mites
August – 33 mites

If in excess of these figures, the colony will collapse before the end of the season. (Northern Hemisphere dates apply here.)

Mite numbers in colony estimation

November–February: daily mite fall × 400
March–April and September–October: daily mite fall × 100
May–August: daily mite fall × 30

(See also Van Eaton and Goodwin, 2001; CAAPE, n.d.)

Chemical control *(miticides)*

Bayvarol, Apistan, Apivar, Check Mite and Apitol are proprietary treatments designed to treat Varroa. They are based on synthetic chemicals. Where no chemical resistance is shown, they are reliable and effective. They are also easy and quick to use (Van Eaton and Goodwin, 2001).

'Organic' chemicals

Thymol, formic acid, lactic acid, oxalic acid and other essential oils are 'soft' chemicals that are reliable to varying degrees but, for good effect, they must be used at the right time and in the right circumstances. Some are very temperature dependent. Commercial treatments are available based on some of these substances, such as Apilife Var (various oils and thymol), Apicure (formic acid) and Apiguard (thymol). Devices for the efficient use of oxalic and lactic acids are also available.

Biotechnical control *(manipulations)*

- **Drone-brood trapping:** if carried out at the correct time, this will dramatically reduce the number of Varroa and will not affect the colony.

- **Worker-brood trapping**: if done correctly, this can be successful but it affects the colony (Van Eaton and Goodwin, 2001).
- **Hive splitting/drone trapping:** this is a very effective way of reducing Varroa numbers (Van Eaton and Goodwin, 2001).
- **Mesh bottom boards:** these are not successful on their own, but they are a help.
- **Heat treatment:** this works but is time-consuming and only 50–80% effective, according to US research.

Co-ordination with neighbours

This helps to reduce re-invasion from untreated stocks and can play a vital part in any Varroa treatment programme.

Biological controls

- **Suppression of mite reproduction (SMR):** this is a trait of mite infertility that is probably related to pupal chemical-trigger release and its propagation by breeding programmes.
- **Breeding programmes:** these mainly involve the selection of traits hostile to the Varroa-mite reproduction cycle or of worker traits that actively remove or hinder the mites.
- **Fungal pathogens:** research has shown that the fungus *Matarhizium anisopliae* may have an effective role in the biological control of Varroa (it is thought to be as effective as a Fluvalinate-based miticide).

SECTION 50:

TROPILAELAPS CLARAE

Identification

- *Tropilaelaps clarae* is smaller than Varroa and elongated, not crab shaped (see Photographs 35 and 36).
- A high infestation shows an irregular, punctured brood pattern and malformed brood. Adult bees may have malformed wings (see also Varroa, Section 49, and foul broods, Sections 41 and 42).
- Similar diagnostic tests to Varroa can be used to determine whether *T. clarae* exists.
- An inspection of capped brood will indicate mite infestation.

Treatment and control

Treatment is similar to that used in Varroa control. *T. clarae* has a major weakness in that it cannot exist outside the cell for long. Brood-less periods will therefore clean the hive of mites.

Pointers

- *T. clarae* will cause colony collapse faster than Varroa.
- It can mate outside cells as well as inside cells.
- It cannot live outside cells during its phoretic period for very long (DEFRA, 2005).

Notes:

- *Although* T. clarae *is not yet thought to be present in Europe or the USA, beekeepers should keep a careful look out for this mite.*
- *The symptoms of* T. clarae *infestation may resemble foul-brood disease.*

35. Tropilaelaps clarae – *elongated*

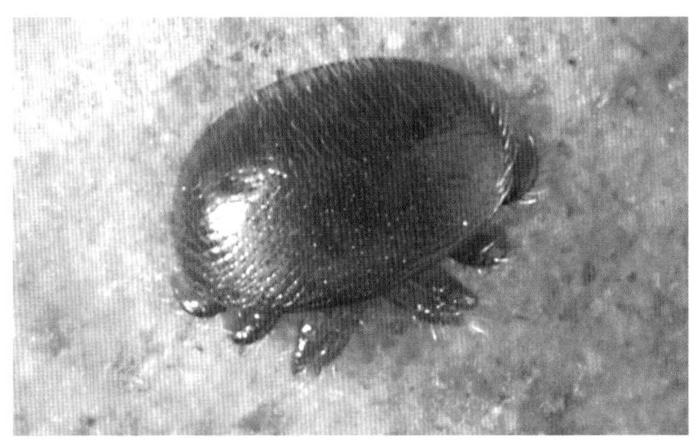

36. *Do not confuse* T. clarae *with Varroa destructor*

SECTION 51:
TRACHEAL MITE (ACARINE)

The mite *Acarapis woodii* (see Photograph 37) invites much controversy when assessing its effects on bees. The mite inhabits the prothoracic trachea of the honeybee – the thoracic opening nearest to the bee's head on the thorax. These openings are really air inlets which allow air to enter the bee's bloodstream. They enter this opening a few days after the bee emerges when the hairs surrounding the opening are still soft. Bees seem to vary in their susceptibility to this mite. In the USA, for example, it is a major problem whereas, in Europe, it is a minor player and causes little damage.

Identification

- There are no certain field methods for determining an infestation of *A. woodii.*
- Dissection of the bee is the only certain way for the beekeeper to determine the presence of this mite
- There are no certain visible symptoms in the field.

Treatment, prevention and control

- Certain evaporative treatments, such as those using menthol or formic acid and which are employed for Varroa control, are also used for the treatment of tracheal mite.
- Treatment is generally unnecessary in Europe.

Pointers

- According to certain researchers (Furgala *et al*., 1989), when over 30% of the bees in a colony become infected, honey production may be reduced, and the chances of the winter survival of the colony decreases with a corresponding increase in infestation.

- Many of the symptoms generally ascribed to Acarine or tracheal mites may be those belonging to viral diseases or PMS (see Section 52).

- This mite was identified (almost certainly wrongly, although it may have had a part to play as a virus vector) as being solely responsible for the Isle of Wight disease of the early twentieth century.

- Its apparent limited effect in Europe and its heavy effect in the USA indicate great scope for further research on this enigmatic mite.

(See also Bailey, 1999.)

37. *Tracheal mite* (Acarapis woodii)

SECTION 52:
PARASITIC MITE SYNDROME (PMS)

PMS is the name given to a range of abnormal brood symptoms associated with the presence of Varroa on both brood and adult bees. The symptoms were found in association with infections of both Varroa and Acarine (tracheal mite) in the USA.

The syndrome affects both brood and adult bees and is usually associated with colony collapse, especially in the autumn. Symptoms can appear at any time of the year, although they are more prevalent in mid-summer and autumn. They are often difficult to interpret and can be very easily confused with the symptoms of AFB, EFB, sac brood and virus diseases, so be careful how you interpret this in the field.

Causes

PMS is thought to be caused by Varroa vectoring the acute bee paralysis virus (ABPV) and possibly other viruses into honeybee larvae (see also Varroa, Section 49; AFB, Section 41; EFB, Section 42; sac brood, Section 44; and virus diseases, Section 48). However, in other cases neither ABPV, Kashmir bee virus (KBV) nor any of nine other bee viruses were found, so it is evident that, while these viruses may be one of the causes of the syndrome, other factors cannot be ruled out.

Identification (see Photograph 38)

Symptoms in the brood

- The symptoms are difficult to interpret. They are similar to those of foul brood (see Sections 41 and 42) and/or sac brood (see Section 44), but there is no ropiness or bad odour. The scales are easy to remove. (If necessary, use a commercial field test for AFB.)

- The affected larvae die in the pre-pupal or late larval stage.
- The larvae are found stretched out in the cell with the head raised (like sac brood).
- Early on, the larvae are dull white in colour and deflated. Later, they become grey with brownish spots.
- The cappings are often perforated or uncapped by the bees (like AFB).
- If stirred with a match, the larva is globular not ropey (like EFB).
- Affected brood can be of any age and anywhere in the comb.

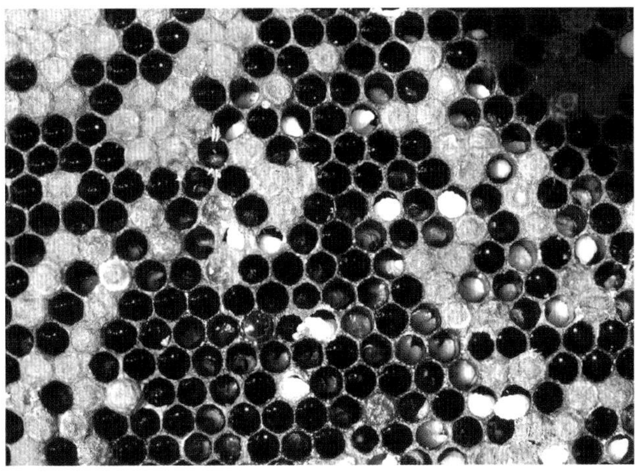

38. *Typical view of suspected PMS*

Adult symptoms *(these are more difficult to determine accurately)*

- Crawling/moribund bees may be seen.
- The adult population is much reduced. This occurs rapidly.

Treatment, prevention and control

- The use of Varroa treatments (see Section 49) suggests a close link to the presence of Varroa.
- Some authorities suggest that feeding sugar syrup benefits the colony.
- If you keep your colonies strong and treated for Varroa, this syndrome should not appear. Do not leave treatment for mites until it is too late.

Pointers

- In all cases, Varroa is associated with this syndrome.
- The pre-pupal stage dies after the cell is capped.
- The larvae are commonly infected with one or more viruses.
- If you do see the symptoms above, check first for AFB or EFB and sac brood.
- Autumn colony collapse is common if the syndrome is not checked. If you wait until the syndrome is advanced, you will not be able to treat the colony successfully.

Note:
This syndrome exhibits symptoms similar to several other diseases and so careful checks should be carried out to ensure that no confusion arises. Commercial AFB/EFB field tests now exist that can be purchased from bee-supply stores.

SECTION 53:

COLONY COLLAPSE DISORDER (CCD)

CCD is a comparatively recent phenomenon that has entered the catalogue of honeybee problems, and it can be considered one of the most devastating, causing huge losses in many countries to the extent that beekeeping and agriculture have been severely affected in parts. First noticed in Spain on a grand scale in 2004, the problem has spread to many parts of the world and is being extensively researched. Despite the fact that it is an immensely complex subject, identification in the field is easy – astonishingly so.[3]

Field identification

In cases where the colony appears to be actively collapsing:

- There is an insufficient workforce to maintain the brood that is present.
- The workforce seems to be made up of young adult bees only.
- The queen is present, appears healthy and is usually still laying eggs.
- The cluster is reluctant to consume feed provided by the beekeeper, such as sugar syrup and protein supplement.
- Foraging populations are greatly reduced/non-existent.

In cases where a colony has collapsed

A colony that has collapsed from CCD is generally characterised by all the following conditions occurring simultaneously:

- The presence of capped brood in abandoned colonies. Bees will not normally abandon a hive until the capped brood has hatched.
- The presence of food stores, both honey and bee pollen, which are not immediately robbed by other bees.
- When attacked by hive pests such as wax moth and small hive beetle, the attack is noticeably delayed.
- The presence of the queen bee. (If the queen is not present, the hive died because it was queenless, which is not considered CCD.)

Causes

The causes of this problem are now thought to be a combination of many factors – not just one or two. Increasing evidence suggests that the use of systemic neonicotinoid insecticides such as imidacloprid may cause increased vulnerability to infection in insects. Research shows that increased disease infection happens even when the levels of the insecticide are so tiny they could not be detected in the bees, (although the researchers knew that they had been dosed with it).

Pathogen levels

A higher total load of pathogens – viruses, bacteria and fungi – appears to have the strongest link with CCD. CCD colonies are co-infected with a greater number of pathogens – bacteria, micro-parasites (such as Nosema) and viruses. Overall, 55%

of CCD colonies were infected with three or more viruses, compared with 28% of non-CCD colonies.

Israeli acute-paralysis virus (IAPV) levels

Research has established a link between a new virus, IAPV, and CCD colonies. Of those colonies that suffered from CCD, all had IAPV, while healthy colonies did not.

Nosema ceranae and viruses

Other research suggests that this link with IAPV may not be accurate and that two other pathogens, when working in concert, can lead to the decline in a honeybee colony and CCD. It found a slightly higher incidence of a fungal pathogen known as *Nosema ceranae* in sick colonies, but it was not statistically significant until it was paired with other pathogens. Levels of the fungus were slightly higher in sick colonies, but the presence of that fungus and two or three RNA viruses from the family *Dicistroviridae* is a pretty strong predictor of collapse.

Summary

In most of the research no single variable was found consistently in only those honeybee colonies that had CCD, and so now the search for factors that are involved in CCD is focusing on four areas: pathogens, parasites, environmental stresses and bee-management stresses, such as poor nutrition. It is unlikely that a single factor is the cause of CCD; it is more likely there is a complex of different components.

SECTION 54:
SMALL HIVE BEETLE (SHB)

The arrival of this pest on the beekeeping scene in the USA and Australia has been a major blow to beekeeping in these countries. The beetle can multiply to huge numbers in a beehive, and it will can eat the brood, destroy the comb and quickly end the life of a colony. Its home is in Africa where it is regarded as a minor pest of honeybees, but its presence outside this area is, like so many pests and diseases that have spread throughout the world, a disaster. Beekeepers in beetle-free areas of the world need to look out for this pest and report any findings immediately. Keep an eye out for any signs in the apiary.

Identification *(see Photographs 39 and 40)*

- The adult beetle is about one third the size of a bee and is readily seen. They are initially reddish brown but mature to black. Adults have two distinctive club-shaped antennae.

39. *Small hive beetle (SHB)*

- When a hive is opened, adult beetles can be seen running across the combs to hide from the light.
- If the infestation is heavy, adults may be seen on hive floors and under lids.
- Small, pearly white eggs, smaller than bee eggs, can be found in irregular masses in crevices or brood combs.

- If you leave corrugated cardboard pieces in the hive, the beetles will be attracted to them for shelter and so are readily found if their presence is suspected.
- The smell of fermented honey (caused by larvae excreting in the honeycomb) is distinctive.
- The larvae are similar in size to wax-moth larvae. After 10–14 days they are 10–11 mm long. They have three pairs of small protolegs near to their heads and spines on their backs.
- The larvae do not produce webbing or frass in the combs, thus differing from wax moth.
- Infested combs have a slimy appearance.

Damage

- Both adults and larvae will eat bee eggs, brood, honey and pollen, thus ultimately destroying the entire colony.
- Colonies often abscond when heavily infested.
- Heavy infestations reach tens of thousands of larvae, causing partial comb meltdown.
- Fermented, spoiled comb is repellent to bees.

Treatment

- Beetle traps may be effective, and some are now on the market in the USA and Australia that have a high degree of success.
- Fluorescent-light, larvae-attracting traps can be used.
- Soil drenches in front of the hive.
- Chemical strips in the hive.

- Good hygiene, especially in the extracting room.
- The most effective control against SHB is maintaining colony strength. Coupled with minimising empty frames of comb, this should all but eliminate the chances of colony failure.

40. *SHB (note the club-shaped antennae and the protolegs on the larva)*

Pointers

- Beetles have been found in swarms.
- It appears that, the day after an apiary inspection occurs, there is a huge influx of beetles. (The hive odours released attract beetles from up to 10 miles away.)
- Opening a hive provokes existing beetles to lay eggs.
- Stored comb in a honey-extraction room is especially at risk.
- The adults and larvae will eat bee eggs, brood, honey and pollen.
- Colonies will often abscond when heavily infested.

(See also MAFF, 2000.)

SECTION 55:

OTHER PESTS OF HONEYBEES

There are many other pests that can make the beekeeper's life difficult, and these include birds, animals and other insects. Most other pests are not significant even though they may have serious local effects. There are various methods of getting round the problems, involving the beekeeper and the local authorities. The following are some of the pests you may see out in the field, depending on which country you are based in:

- **Large animals:** these include cows and horses. The solution, of course, is to ensure that your apiary is well wired off, with electrical tape if necessary.
- **Bears:** these can be a serious local pest but, in areas where they operate, there are strategies for dealing with them (e.g. in Spain they are encouraged to ravage specially set-up apiaries in bear areas). This is more to preserve the few bears that are left than to preserve the bees, although the system works well to the satisfaction of the beekeepers and the bears.
- **Skunks:** these are a local pest in the USA and have the effect of making the bees aggressive.
- **Bee eaters:** these are a pest in several areas of Europe and Africa, but there is a great deal of evidence to suggest that their significance as a pest is low and that they eat many bee enemies as well.
- **Herons:** herons have been known to attack beehives in cold winters, even to the extent of destroying the hive walls to get access to the bees. This is not a significant problem and, in

rare cases where it does occur, the solution is to put garden netting over the hives.

- **Ants:** these again are a local pest in many areas. In Europe they tend to signal that a hive has problems and that the colony has disappeared or is suffering from disease. Healthy colonies can usually keep them at bay.

- **Wasps:** these are a pest, especially later in the year. The solution is to reduce hive entrances and to set up beer-bottle traps near to hive entrances. It is sometimes necessary to follow the wasps to their nest and to destroy this.

- **Mandarin hornets:** this is more a pest in the East than in Europe or the USA. This beast can decimate a colony in no time. It simply sits there, decapitating the guard bees, and then it goes for the brood. However fascinating this may be, it would obviously go against your management strategy. The eastern honeybee has sorted this problem out by balling the hornet with a group of bees and raising the temperature within the ball of bees to a degree more than the hornet can stand. The bees can survive this temperature but the hornet dies. The western honeybee, *Apis mellifera*, doesn't know this trick, so watch out for it and beware.

None of the above problems should be dismissed but, instead, solutions should be found that enable beekeepers to maintain healthy apiaries and, hopefully, wildlife to maintain their numbers.

Author's notes

1. *I have never seen this in any colony, although I have seen enough wax-moth damage to last a lifetime.*

2. *I used PDB crystals once. I found out about* Bacillus thuringiensis *and never looked back. It is natural and does not contaminate anything except the wax-moth larvae. PDB is now banned from use in many countries due to research showing carcinogen links. It can cause mould in the comb, though – especially in stored comb.*

3. *My bees in Spain were perhaps the first to suffer from CCD in 2004 – or perhaps I was the first to admit I had a problem. It was an eerie sensation seeing strong hives reduced to nothing in a matter of days. My neighbours just wrote me off as a bad beekeeper, but little did I know then that many of them were also suffering from the problem and not admitting it or blaming it on a host of other causative agents.*

Part I

A beekeeper's ready-reckoner

Introduction

This part is full of beekeeping facts and figures:

- When will the queen emerge?
- How can I make queen candy?
- How much feed do my bees need for winter and how can I work it out?
- When will my queen mate?

And so on. This part attempts to give rapid field answers to many questions that necessarily pop up and that you should know the answer to and thought you did – until you find yourself out in the bee shed, miles from your books with bags of sugar and yeast to mix! Important information is also given on Varroa mite numbers and counts for your all-important Varroa monitoring programme.

SECTION 56:
QUEEN/WORKER/DRONE DEVELOPMENT

As a reminder, Figure 8 shows the average development times of queens, workers and drones (see the notes below):

Figure 8. Queen/worker/drone development

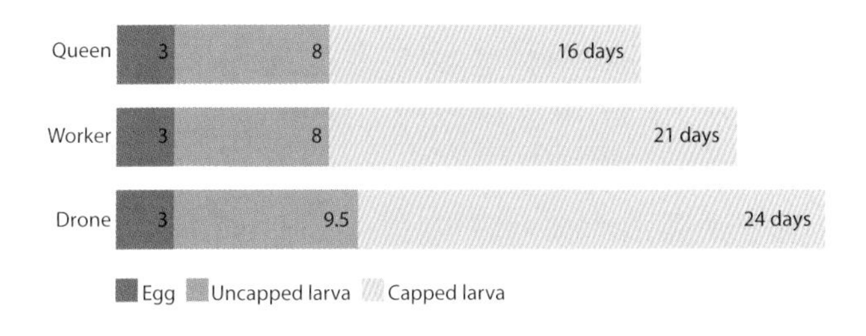

- **Queen on emergence:** if the climatic conditions permit, the queen will make a mating flight around 5–6 days after emergence. She will start to lay eggs 36 hours or more after a successful flight (more usually after 3 days).
- **Drone on emergence:** fed by workers until around 7 days old. Remains in the hive until approximately 12–13 days old (sexually mature). Thereafter the drone will take mating flights during the afternoons. They are often kicked out of the hive during the autumn or during other times of dearth.
- **Worker on emergence:** this is a complex subject that is not appropriate for a short field guide (see Schmidt, 2001; CAAPE, n.d.). A worker's lifespan will vary according to the time of

year. During the summer, the average life span is 15–38 days; during the winter it can be 140 days or more. This depends very much on the conditions prevailing at the time. Foraging worker bees in the summer literally work themselves to death whereas, in winter, they will be left alone doing very little in the hive.

Note:
The number of days until emergence for each caste can vary considerably (e.g. queen: 14–17 days; worker: 16–24 days; drone: 20–28 days). This variability may be due to environmental factors (especially temperature) and nutrition and, again, is the subject of much research (Bailey, 1999).

SECTION 57:
FEED MIXES

The following mixes have been shown to be effective for bees and are used by many, very large-scale commercial beekeepers. Hobbyist beekeepers should just reduce the quantities to manageable proportions.

Sugar syrup

- Thick feed for autumn feeding: 1 kg sugar to 500 ml water.
- Thin feed for spring stimulation or pollination feeding: 1 kg sugar to 1 l water.

Invert sugar syrup

The following amount is for large quantities:

- 1,000 l sugar syrup at 30–40 °C.
- 250 gm dried, active baker's yeast.
- 1 l warm water.

1. Mix the yeast with a cup of sugar syrup and the litre of warm water (at around 35–40 °C).
2. When it starts to rise, pour the mix into a 1,000-l vat of sugar syrup and stir well.
3. Increase the temperature to 65 °C ensuring that it remains for at least 2 hours between 45 °C and 55 °C. Once the temperature reaches 65 °C, turn off the heat and allow to cool.

Queen candy

This recipe can use bulk-purchased sugar syrup or your own homemade syrup. It makes sufficient candy for around 350 queen cages:

- Sugar syrup made from 2 cups of white sugar to 1 cup of water.
- 2 kg icing sugar.
- ¼ teaspoon tartaric acid.
- 2 teaspoons glycerine.

Or, more expensively, honey and icing sugar mixed to a stiff paste. This is more difficult to maintain as a firm mixture and tends to melt in warmer weather, thus releasing the queen too early if employed as a plug in a queen cage.

Pollen substitute

When used at the right time, pollen substitutes can be a vital supplement for colonies. Start feeding them about 4–5 weeks before brood rearing commences and keep feeding until natural pollen is plentiful. Mix the following ingredients:

- 1 part sodium caseinate (a readily available dairy derivative).
- 2 parts non-active dried yeast.
- Sugar syrup to make a stiff paste. (Ensure that the sugar syrup is not fermenting, otherwise the patties will blow up.)

Mix in a cake or commercial baker's blender if large quantities are being made. Fill small paper bags with the mix and, when you give them to the bees, open the upper side of the pattie bags.

Note:
Avoid the use of soya protein in bee feed.[1]

SECTION 58:
FEED FACTS

This section provides information for preparing feed for bees and for calculating winter feed requirements. It is important that you know exactly what stores your bees have, especially before going into winter, and then to feed if necessary.

Sugar syrup

Sugar syrup will ferment easily if wild yeast enters it or if baker's yeast is not killed off following inversion. Table 15 gives the temperatures and times needed to kill sugar-tolerant yeasts.

TABLE 15: Temperatures and times to kill sugar-tolerant yeasts

Degrees (°C)	Time (mins)
51.7	470
54.4	170
57.2	60
60.0	22
62.8	7.5
65.6	2.8
68.3	1.0

Feeding calculation measurements

Each 5 l of heavy syrup will increase stores by 3 kg. In 4.5 kg of honey there is 3.5 kg of sugar so, if the colony is 4.5 kg short, feed 3.5 kg of sugar. For other shortages, therefore, multiply the shortfall by 0.8.

When working out how much stores your bees have, use the following approximate weights of honey in the comb (figures rounded up):

- Each Langstroth frame contains approx. 3 kg honey.
- Each shallow Langstroth frame contains approx. 1.9 kg honey.
- Each BS frame contains 2.5 kg honey.
- Each BS shallow frame contains approx. 1.6 kg honey.
- Each MD deep frame contains approx. 3.9 kg honey.
- Each full ¾ Langstroth frame covered with bees has approx. 2,300 bees.
- Each full Langstroth frame covered with bees has approx. 3,500 bees.

Note:
All the above figures concerning comb weights are approximate. For example, a rare super containing perfect combs correctly bee spaced and totally sealed will contain a greater weight of honey. Poorly built combs may contain less.

SECTION 59:
WEIGHTS AND MEASURES

Many of the figures in this Field Guide and in other beekeeping books may use measurements you are unused to (e.g. metric, US or imperial). The conversions below may be of help. Quantities have been rounded up and are for field use only. For more exact measurements you are advised to consult a dedicated weights and measures ready-reckoner or one of the many online conversion sites.

Abbreviations:

oz = ounce
lb = pound
cwt = hundredweight

pt = pint
gal = gallon

in = inch
ft = foot (feet)
yd = yard
nm = nautical mile

ml = millilitre

l = litre

g = gram
kg = kilogram

mm = millimetre
cm = centimetre
m = metre
km = kilometre

ha = hectare

fl = fluid
sq = square

Linear measurement:

1 in = 25 mm
1 ft = 305 mm (0.3 m)
1 yard = 915 mm (0.9 m)
1 mile = 1.6 km

1 cm = $\frac{3}{8}$ in
1 m = 3 ft 3 in
1 km = 0.6 mile
1 nm = 1.8 km (legally, but not by computation)

Area:

1 sq in = 6.5 cm^2

1 sq ft = 0.09 m^2

1 sq yd = 0.8 m^2

1 sq mile = 2.6 km^2

1 acre = 0.4 ha

1 ha = 2.5 acres

Liquids:

1 pt = 0.6 l

1 UK gal = 1.2 US gal = 4.5 l

1 US gal = 0.8 UK gal = 3.8 l

1 fl oz = 28 ml

1 l = 1.8 pt = 4.2 cups (US)

Liquid weights:

1 pt water = 1.3 lb = 0.6 kg

1 gal water = 10 lb = 4.5 kg

1 pt = 20 fl oz = 570 ml

1 l water = 1 kg = 2.2 lb

Weights:

1 oz = 28.4 g

1 lb = 0.5 kg

1 UK cwt = 1.1 US cwt = 50.8 kg = 112 lb

1 UK ton = 1 metric tonne = 1.12 US ton = 20 cwt

1 kg = 2.2 lb

Temperature:

To convert Fahrenheit to centigrade (Celsius):

x °F = (x – 32) ÷ 1.8 (e.g. 60 °F = (60 – 32) ÷ 1.8 = 15.5 °C)

To convert centigrade (Celsius) to Fahrenheit:

y °C = (y × 1.8) + 32 (e.g. 60 °C = (60 × 1.8) + 32 = 140 °F)

SECTION 60:
VARROA-MITE NUMBER CALCULATIONS

When you sample the whole hive for Varroa numbers, usually only 15% (approx.) of mites will be on adult bees in a hive in full production. A correction factor is therefore needed to account for the rest:

1. Using a clean sticky board, place Apistan or Bayvarol as normal.
2. After 24 hours count the mites. Assume a 85% kill rate.
3. Divide the number of mites counted by 0.85.
4. If the hive is in full production, multiply the result by 6.
5. If the hive is not in full production but has brood, multiply by 3.
6. If no brood is present, then no correction factor is required.

Ether-roll field test to determine Varroa numbers:

1. Collect about 3–500 bees in a jar.
2. Spray ether into the jar for 1–2 seconds (use a can of quick start).
3. Rotate the jar for about 10 seconds.
4. Deposit the bees on to a white piece of paper. Do this immediately after the 10 seconds.

5. Spread the bees around to dislodge more mites.
6. Count the mites. In this example let the number of mites seen be 5.
7. Divide the number of bees in the hive by the number in the sample: 25,000 in hive ÷ 500 in sample = 50.
8. Multiply 50 by 5 (number of mites seen) = 250.
9. Then multiply by 6 if the hive is in full production = 1,500.
10. *OR* by 2 if not in production but contains brood = 500.
11. *OR* by 0 if no brood is present = 250.

Field test for mite resistance to chemical-based miticides:

1. Cut a 9 mm × 25 mm piece of Apistan strip. Staple it to a 125 mm × 75 mm piece of card.
2. Place the card in a 500-ml glass jar.
3. Prepare a light metal-mesh cover for the jar.
4. Shake bees from 1–2 combs into an upturned hive roof.
5. Scoop about a quarter of these (about 150) and place in a jar with a sugar cube.
6. Cover with the mesh lid and store upturned in the dark at room temperature.
7. After 24 hours, place the upturned jar over some white paper and hit it to dislodge dead mites. Count the initial mite kill.

8. Put the bees in the freezer until the bees are dead (1–4 hours).
9. Take out and count the final mite kill.
10. Percent kill by the chemical strip = Initial kill ÷ (Initial + Final kill) × 100.
11. If less than 50%, the probability of resistance exists.

Note:
Fine sugar or soapy water can also be used in the roll test instead of ether.

SECTION 61:
POLLINATION REQUIREMENTS

The pollination of crops is one of the most important jobs of the honeybee. It's what bees are all about and, in a world where natural habitat and native pollinating insects are rapidly disappearing, the honeybee's role is becoming ever more important. In the modern world of monocrop cultivation, some crops require a very high level of pollinators to produce economically viable crops. This may be because of the blossoms' lack of attractiveness or because alternative pollinators are employed when the native pollinators are extinct or rare.

Crops such as alfalfa, cranberries and kiwi fruit all demand special attention. Kiwi fruit, for example, do not give nectar, only pollen, and so the bees must be introduced at exactly the right time when the flowers are at 10% bloom. Any earlier and the bees may be diverted to more attractive crops or wildflowers in the area and, any later, the orchardist won't get value for money. For many crops, the requirement for hives per acre is rising because there are no feral bees (Varroa) and because specialist pollinators don't exist any more. Some crops are not efficiently pollinated by honeybees (such as alfalfa (lucerne)) but, by trucking in huge numbers of hives, their deficiencies can be overcome.

Because of different climatic and other conditions around the world, there is a wide variety of hive number estimations for different crops. Table 16 shows the generally agreed average number of hives required per acre and hectare. Find out the best averages for your area. Farmers and orchardists will err on the side of bigger numbers, whereas the beekeeper will do the opposite.

Table 16: Average number of hives required (per acre and hectare)

Crop	Per acre	Per hectare
Alfalfa	1 (3–5)	2.5 (4.9–12)
Almonds	2–3	4.9–7.4
Apples (dwarf)	3	7.4
Apples (normal size)	1	2.5
Apples (semi-dwarf)	2	4.9
Apricots	1	2.5
Blueberries	3–4	7.4–9.9
Borage	0.6–1	1.5–2.5
Buckwheat	0.5–1	1.2–2.5
Canola (hybrid)	2.0–2.5	4.9–6.2
Canola	1	2.5
Cantaloupes	2–4 (average 2.4)	4.9–9.9 (average 5.9)
Clovers	1–2	2.5–4.9
Cranberries	3	7.4
Cucumbers	1–2 (average 2.1)	2.5–4.9 (average 5.2)
Ginseng	1	2.5
Musk melon	1–3	2.5–7.4
Nectarines	1	2.5
Peaches	1	2.5
Pears	1	2.5
Plums	1	2.5
Pumpkins	1	2.5
Raspberries	0.7–1.3	1.7–3.2
Squash	1–3	2.5–7.4
Strawberries	1–3.5	2.5–8.6
Sunflower	1	2.5
Trefoil	0.6–1.5	1.5–3.7
Watermelon	1–3 (average 1.3)	2.5–4.9 (average 3.2)
Zucchini	1	2.5

Author's note

1. *I have read research that suggests soya has a deleterious effect on the queen's ovaries but I am unable to find this research again and so put it in just for advice.*

Part I

Part J

Introduction

This short part is a guide to further information and reading. Bees and beekeeping is a vast subject. I have kept the references to a minimum and have tried to refer only to those more readable papers and articles on the relevant subject.

The glossary of terms provides a list of common beekeeping terms used throughout the English-speaking world and does not necessarily include all the words used in the Field Guide.

There are many good books on beekeeping which, together, cover virtually every aspect of the craft and science. I have listed only a few books in the further reading section which, in my opinion, are classic 'must have' books.

Beekeeping organisations come and go and so probably the best guide to these is published as the *Beekeeper's Annual* by Northern Bee Books. I have only entered the main organisations that tend to have a global presence and a website.

SECTION 62:

GLOSSARY OF TERMS

Remember that terms may differ in various regions and countries that use English, so some of the terms will be unfamiliar to you. This is not necessarily a list of beekeeping words used in this Field Guide but the list should enable you to refer to unfamiliar words from other texts.

Abdomen: the posterior or third region of the body of a bee enclosing the honey stomach, true stomach, intestine, sting and reproductive organs.

Absconding swarm: an entire colony of bees that abandons the hive because of disease, wax moth or other maladies.

Adulterated honey: any product labelled 'Honey' or 'Pure Honey' that contains ingredients other than honey but does not show these on the label.

After-swarm: a small swarm, usually headed by a virgin queen, which may leave the hive after the first or prime swarm has departed.

Alighting board: a small projection or platform at the entrance of the hive.

American foul brood (AFB): a brood disease of honeybees caused by the spore-forming bacterium, *Paenibacillus larvae*.

Anaphylactic shock: a severe allergic reaction that can be brought about by bee venom, causing a constriction of the muscles surrounding the bronchial tubes of a human. It is caused by hypersensitivity to venom and can result in death unless immediate medical attention is received.

Apiary: colonies, hives and other equipment assembled in one location for beekeeping operations (USA 'bee yard').

Apiculture: the science and art of keeping honeybees.
Apis mellifera: scientific name of the honeybee found in most of Europe and the USA.
Automatic uncapper: automated device that removes the cappings from honeycombs, usually by moving heated knives, metal teeth or flails.

Bait hive: a hive set up to attract swarms.
Bee blower: an engine with attached blower used to dislodge bees from combs in a honey super by creating a high-velocity, high-volume wind.
Bee bread: a mixture of collected pollen and nectar or honey, deposited in the cells of a comb to be used as food by the bees.
Bee brush: a brush used to remove bees from combs. Can be in the form of a feather.
Bee escape: a device used to remove bees from honey supers and buildings by permitting bees to pass one way but preventing their return.
Bee metamorphosis: the three stages through which a bee passes before reaching maturity: egg, larva and pupa.
Bee space: 60mm –1 cm space between combs and hive parts in which bees build no comb or deposit only a small amount of propolis.
Bee veil: a cloth or wire netting for protecting the beekeeper's head and neck from stings. Now usually part of the full bee suit.
Bee venom: the poison secreted by special glands attached to the sting of the bee.
Beehive: a box or receptacle with movable frames, used for housing a colony of bees.
Beeswax: a complex mixture of organic compounds secreted

by special glands on the last four visible segments on the ventral side of the worker bee's abdomen and used for building comb.

Benzaldehyde: a volatile, almond-smelling chemical used to drive bees out of honey supers. Sold under trade names such as 'Bee Go'.

Boardman feeder: a device for feeding bees in warm weather, consisting of an inverted jar with an attachment allowing access to the hive entrance. Not as useful as a frame feeder.

Bottom board: the floor of a beehive.

Brace comb: a bit of comb built between two combs to fasten them together, between a comb and adjacent wood, or between two wooden parts, such as top bars.

Braula coeca: the scientific name of a wingless fly commonly known as the bee louse. This small fly is rarely a problem and often goes unnoticed by beekeepers.

Brood: bees not yet emerged from their cells: eggs, larvae and pupae.

Brood chamber: the part of the hive in which the brood is reared; may include one or more hive bodies and the combs within.

Buff comb: a bit of wax built on a comb or on a wooden part in a hive but not connected to any other part.

Capped brood: pupae whose cells have been sealed with a porous cover by mature bees to isolate them during their non-feeding pupal period; also called sealed brood.

Cappings: the thin wax covering of cells full of honey; the cell coverings after they are sliced from the surface of a honey-filled comb.

Castes: the three types of bees that comprise the adult population of a honeybees colony: workers, drones and queen.

Cell: the hexagonal compartment of a honeycomb.

Cell bar: a wooden strip on which queen cups are placed for rearing queen bees.

Cell cup: base of an artificial queen cell, made of beeswax or plastic and used for rearing queen bees.

Chilled brood: immature bees that have died from exposure to cold; commonly caused by mismanagement.

Chunk honey: honey cut from frames and placed in jars along with liquid honey.

Clarifying: removing visible foreign material from honey or wax to increase its purity (fine filtering).

Cluster: a large group of bees hanging together, one upon another.

Colony: the aggregate of worker bees, drones, queen and developing brood living together as a family unit in a hive or other dwelling.

Comb: a wax slab of six-sided cells made by honeybees in which brood is reared and honey and pollen are stored; composed of two layers united at their bases.

Comb foundation: commercially made wax sheets consisting of thin sheets of beeswax with the cell bases of worker cells embossed on both sides in the same manner as they are produced naturally by honeybees.

Comb honey: honey produced and sold in the comb, in either thin wooden sections or circular plastic frames.

Creamed honey: honey which has been allowed to crystallise, usually under controlled conditions, to produce a tiny crystal.

Crimp-wired foundation: comb foundation into which crimp wire is embedded vertically during foundation manufacture.

Cross-pollination: the transfer of pollen from an anther of one plant to the stigma of a different plant of the same species.

Crown board: see 'Inner cover'.

Crystallisation: see 'Granulation'.

Cut-comb honey: comb honey cut into various sizes, the edges drained and the pieces wrapped or packed individually.

Decoy hive: a hive placed to attract stray swarms (see also 'Bait hive').

Demaree: the method of swarm control that separates the queen from most of the brood within the same hive.

De-queen: to remove a queen from a colony.

Dextrose: one of the two principal sugars found in honey; forms crystals during granulation. Also known as glucose.

Dividing: separating a colony to form two or more units.

Division board feeder: a wooden or plastic compartment which is hung in a hive like a frame and contains sugar syrup to feed bees.

Double screen: a wooden frame, with two layers of wire screen to separate two colonies within the same hive, one above the other. An entrance is cut on the upper side and placed to the rear of the hive for the upper colony.

Drawn combs: combs with cells built out by honeybees from a sheet of foundation.

Drifting of bees: the failure of bees to return to their own hive in an apiary containing many colonies. Young bees tend to drift more than older bees and bees from small colonies tend to drift into larger colonies.

Drone: the male honeybee.

Drone comb: comb measuring about four cells per linear 2.5 cm that is used for drone rearing and honey storage.

Drone layer: an infertile or unmated laying queen.

Drumming: pounding on the sides of a hive to make the bees ascend into another hive placed over it.

Dwindling: the rapid dying off of old bees in the spring; sometimes called spring dwindling or disappearing disease.

Dysentery: an abnormal condition of adult bees characterised by severe diarrhoea and usually caused by starvation, low-quality food and moist surroundings.

Electric embedder: a device allowing rapid embedding of wires in foundation with electrically produced heat.

European foul brood (EFB): an infectious brood disease of honeybees caused by streptococcus pluton.

Extracted honey: honey removed from the comb by centrifugal force.

Fermentation: a chemical breakdown of honey, caused by sugar-tolerant yeast and associated with honey having a high moisture content.

Fertile queen: a queen, inseminated instrumentally or mated with a drone, which can lay fertilised eggs.

Field bees: worker bees at least three weeks old that work in the field to collect nectar, pollen, water and propolis.

Flash heater: a device for heating honey very rapidly to prevent it from being damaged by sustained periods of high temperature.

Follower board: a thin board used in place of a frame usually when there are fewer than the normal number of frames in a hive.

Food chamber: a hive body filled with honey for winter stores.

Frame: four pieces of wood designed to hold honeycomb, consisting of a top bar, a bottom bar and two end bars.

Fructose: the predominant simple sugar found in honey; also known as levulose.

Fume board: a rectangular frame, the size of a super, covered with an absorbent material such as burlap, on which is placed a chemical repellent to drive the bees out of supers for honey removal.

Fumidil-B: the trade name for Fumagillin, an antibiotic used in the prevention and suppression of Nosema disease.

Glucose: see 'Dextrose'.

Grafting: removing a worker larva from its cell and placing it in an artificial queen cup in order to have it reared into a queen. Also called larval transfer.

Grafting tool: a needle or probe used for transferring larvae in grafting of queen cells.

Granulation: the formation of sugar (dextrose) crystals in honey.

Hive: an artificial home for bees.

Hive body: a wooden box which encloses the frames.

Hive stand: a structure that supports the hive.

Hive tool: a sturdy metal device used to open hives, pry frames apart and scrape wax and propolis from the hive parts.

Honey: a sweet viscid material produced by bees from the nectar of flowers, composed largely of a mixture of dextrose and levulose dissolved in about 17% water; contains small amounts of sucrose, mineral matter, vitamins, proteins and enzymes.

Honey extractor: a machine which removes honey from the cells of comb by centrifugal force. Generally two variants: 'tangential' and 'radial'.

Honey flow: a time when nectar is plentiful and bees produce and store surplus honey.

Honey gate: a tap or faucet used for drawing honey from drums, cans or extractors.

Honey house: building used for extracting honey and storing equipment.

Honey pump: a pump used to transfer honey from a sump or extractor to a holding tank or strainer.

Honey stomach: an organ in the abdomen of the honeybee used for carrying nectar, honey or water.

Honey sump: a clarifying tank between the extractor and honey pump for removing the coarser particles of comb introduced during extraction.

Honeydew: a sweet liquid excreted by aphids, leafhoppers and some scale insects that is collected by bees, especially in the absence of a good source of nectar.

Increase: to add to the number of colonies, usually by dividing those on hand.

Inner cover: a lightweight cover used under a standard telescoping cover on a beehive (see also 'Crown board').

Instrumental insemination: the introduction of drone spermatozoa into the genital organs of a virgin queen by means of special instruments (sometimes known as artificial insemination).

Invertase: an enzyme produced by the honeybee which helps to transform sucrose to dextrose and levulose.

Larva (plural, larvae): the second stage of bee metamorphosis; a white, legless, grub-like insect.

Laying worker: a worker which lays infertile eggs, producing only drones, usually in colonies that are hopelessly queenless.

Levulose: see 'Fructose'.

Mating flight: the flight taken by a virgin queen while she mates on the wing with several drones.

Mead: honey wine.

Melissococcus pluton: causative bacterial agent of European foul brood (EFB).

Migratory beekeeping: the moving of colonies of bees from one locality to another during a single season to take advantage of two or more honey flows.

Nectar: a sweet liquid secreted by the nectaries of plants; the raw product of honey.

Nectar guide: colour marks on flowers believed to direct insects to nectar sources.

Nectaries: the organs of plants which secrete nectar, located within the flower (floral nectaries) or on other portions of the plant (extra-floral nectaries).

Nosema: a disease of the adult honeybee caused by the fungus *Nosema apis* or *N. ceranae*.

Nucleus (plural, nuclei): a small hive of bees, usually covering from two to five frames of comb and used primarily for starting new colonies, rearing or storing queens; also called a 'nuc.'

Nurse bees: young bees, 3–10 days old, which feed and take care of developing brood.

Observation hive: a hive made largely of glass or clear plastic to permit observation of bees at work.

Out-apiary: an apiary situated away from the home of the beekeeper.

Oxytetracycline: see 'Terramycin'.

Package bees: a quantity of adult bees 1–3 kg, with or without a queen, contained in a screened shipping cage.

Paenibacillus larvae: the bacterium that causes American foul brood.

Paralysis: a virus disease of adult bees which affects their ability to use legs or wings normally.

Parthenogenesis: the development of young from unfertilised eggs. In honeybees the unfertilised eggs produce drones.

PDB (paradichlorobenzene): crystals used to fumigate combs against wax moth (keep away from it).

Piping: a series of sounds made by a queen, frequently before she emerges from her cell.

Play flight: short flight taken in front of or near the hive to acquaint young bees with their immediate surroundings; sometimes mistaken for robbing or preparation for swarming.

Pollen: the male reproductive cell bodies produced by anthers of flowers, collected and used by honeybees as their source of protein.

Pollen basket: a flattened depression surrounded by curved spines or hairs, located on the outer surface of the bee's hind legs and adapted for carrying pollen gathered from flowers or propolis to the hive.

Pollen cakes: moist mixtures of either pollen supplements or substitutes fed to the bees in early spring to stimulate brood rearing.

Pollen insert: a device inserted in the entrance of a colony into which hand-collected pollen is placed. As the bees leave the hive and pass through the trap, some of the pollen adheres to their bodies and is carried to the blossom, resulting in cross-pollination.

Pollen patties: see 'Pollen cakes'.

Pollen substitute: any material such as powdered skim milk, brewer's yeast or a mixture of these used in place of pollen to stimulate brood rearing.

Pollen supplement: a mixture of pollen and pollen substitutes used to stimulate brood rearing in periods of pollen shortage.

Pollen trap: a device for removing pollen loads from the pollen baskets of incoming bees.

Pollination: the transfer of pollen from the anthers to the stigma of flowers.

Pollinator: the agent that transfers pollen from an anther to a stigma: wind, bees, flies, beetles, etc.

Polliniser: the plant source of pollen used for pollination.

Prime swarm: the first swarm to leave the parent colony, usually with the old queen.

Proboscis: the mouthparts of the bee that form the sucking tube or tongue.

Propolis: sap or resinous materials collected from trees or plants by bees and used to strengthen the comb, close up cracks, etc.; also called bee glue.

Pupa: the third stage in the development of the honeybee, during which the organs of the larva are replaced by those that will be used by an adult.

Queen: a fully developed female bee, larger and longer than a worker bee.

Queen cage: a small cage in which a queen and three or four worker bees may be confined for shipping and/or introduction into a colony.

Queen cage candy: candy made by kneading powdered sugar with invert sugar syrup until it forms a stiff dough; used as food in queen cages.

Queen cell: a special elongated cell, resembling a peanut shell, in which the queen is reared. It is usually 2.5 cm or more long, has an inside diameter of about 8 mm and hangs down from the comb in a vertical position.

Queen clipping: removing a portion of one or both front wings of a queen to prevent her from flying.

Queen cup: a cup-shaped cell made of beeswax or plastic which hangs vertically in a hive and which may become a queen cell if an egg or larva is placed in it and bees add wax to it.

Queen excluder: metal or plastic device with spaces that permit the passage of workers but restrict the movement of drones and queens to a specific part of the hive.

Queen substance: pheromone material secreted from glands in the queen bee and transmitted throughout the colony by workers to alert other workers of the queen's presence.

Rabbet: a narrow piece of folded metal fastened to the inside upper end of the hive body from which the frames are suspended.

Rendering wax: the process of melting combs and cappings and removing refuse from the wax.

Robbing: stealing of nectar, or honey, by bees from other colonies.

Royal jelly: a highly nutritious glandular secretion of young bees, used to feed the queen and young brood.

Sac brood: a brood disease of honeybees caused by a virus.

Scout bees: worker bees searching for a new source of pollen, nectar, propolis, water or a new home for a swarm of bees.

Sealed brood: see 'Capped brood'.

Self-pollination: the transfer of pollen from anther to stigma of the same plant.

Self-spacing frames: frames constructed so that they are a bee space apart when pushed together in a hive body.

Skep: a beehive made of twisted straw without movable frames.

Slatted rack: a wooden rack that fits between the bottom board and hive body. Bees make better use of the lower brood chamber with increased brood rearing, less comb gnawing and less congestion at the front entrance.

Slumgum: the refuse from melted comb and cappings after the wax has been rendered or removed.

Smoker: a device in which sacking, wood shavings, cardboard or other materials are burned to produce smoke which is used to subdue bees.

Solar wax extractor: a glass-covered insulated box used to melt wax from combs and cappings by the heat of the sun.

Spermatheca: a special organ of the queen in which the sperm of the drone is stored.

Spur embedder: a device used for mechanically embedding wires into foundation by employing hand pressure.

Sting: the modified ovipositor of a worker honeybee used as a weapon of offence.

Sucrose: principal sugar found in nectar.

Super: any hive body used for the storage of surplus honey. Normally it is placed over or above the brood chamber.

Supersedure: a natural replacement of an established queen by a daughter in the same hive. Shortly after the young queen commences to lay eggs, the old queen disappears.

Surplus honey: honey removed from the hive which exceeds that needed by bees for their own use.

Swarm: the aggregate of worker bees, drones and usually the old queen that leaves the parent colony to establish a new colony.

Swarming: the natural method of propagation of the honeybee colony.

Swarm cell: queen cells usually found on the bottom of the combs before swarming.

Terramycin: an antibiotic used to prevent American and European foul brood (see also 'Oxytetracycline').

Tested queen: a queen whose progeny shows she has mated with a drone of her own race and has other qualities which would make her a good colony mother.

Thin super foundation: a comb foundation used for comb honey or chunk honey production which is thinner than that used for brood rearing.

Transferring: the process of changing bees and combs from common boxes to movable frame hives.

Travel stain: the dark discoloration on the surface of comb honey left on the hive for some time, caused by bees tracking propolis over the surface.

Uncapping knife: a knife used to shave or remove the cappings from combs of sealed honey prior to extraction; usually heated by steam or electricity, or it could just be a serrated bread knife.

Uniting: combining two or more colonies to form a larger colony.

Venom allergy: a condition in which a person, when stung, may experience a variety of symptoms ranging from a mild rash or itchiness to anaphylactic shock. A person who is stung and experiences abnormal symptoms should consult a doctor before working bees again.

Venom hypersensitivity: a condition in which a person, if stung, is likely to experience an aphylactic shock. A person with this condition should carry an emergency insect sting kit at all times during warm weather.

Virgin queen: an unmated queen.

Wax glands: the eight glands that secrete beeswax; located in pairs on the last four visible ventral abdominal segments.

Wax moth: larvae of the moths *Galleria mellonella* or *Achroia grisella*, which seriously damage brood and empty combs.

Winter cluster: the arrangement of adult bees within the hive during winter.

Worker bee: a female bee whose reproductive organs are undeveloped. Worker bees do all the work in the colony except for laying fertile eggs.

Worker comb: comb measuring about five cells to 2.5 cm, in which workers are reared and honey and pollen are stored.

SECTION 63:
REFERENCES

Adam, Brother (1975) *Beekeeping at Buckfast Abbey*. Mytholmroyd, West Yorks: Northern Bee Books.

Allen, M. and Ball, B. (1996) 'The incidence and world distribution of honey bee viruses,' *Bee World*, 77: 141–62.

Ambrose, J.T. (1992) 'Two queen management system', in *The Hive and the Honeybee*. Hamilton, IL: Dadant & Sons.

Bailey, L. (1999) 'The century of *Acarapis woodii*', *American Bee Journal*, 139: 541–2.

Bailey, L. and Ball, B.V. (1991) *Honey Bee Pathology* (2nd edn). London and San Diego, CA: Academic Press.

Bosch, R. and Rothenbuhler, W.C. (1974) 'Defensive behaviour and production of alarm pheromone in honeybees', *Journal of Apicultural Research*, 3: 217–21.

Breed, M.D., Robinson, G.E. and Page, R.E. Jr (1990) 'Division of labour during honeybee colony defence', *Behavioural Ecology and Sociobiology*, 27: 395–402.

Brown, R. (1980) *A simple two queen management system*. Mytholmroyd, West Yorks: Northern Bee Books.

CAAPE (n.d.) *Ensayos preliminaries de Apilife Var contra* Varroa jacobsonii *Oud*. Cordoba: Centro Andaluz de Apicultura Ecologico, Cordoba University.

Cook, V. (1986) *Queen Rearing Simplified*. Mytholmroyd, West Yorks: Northern Bee Books.

Crane, E. and Walker, P. (1983) *Impact of Pest Management on Bees and Pollination*. International Bee Research Association: Cardiff.

CSL UK website (**www.csl.gov.uk/science/organ/environ/bee/diseases/adultdiseases/**).

DEFRA (2005) Tropilaelaps clarae: *Parasitic Mite of Honey Bees*. DEFRA: London.
Delaplane, K.S. (1997) 'Colony growth and swarm management', *American Bee Journal*, 137: 197–99.
Delaplane, K.S. (1998a) 'American foulbrood and its control', *American Bee Journal*, 138: 431–2.
Delaplane, K.S. (1998b) 'Varroa control. Timing is everything,' *American Bee Journal*, 138: 575–6.
Flottum, K. and Morse, R. (eds) (1997) *Honey Bee Pests, Predators and Diseases* (3rd edn). Medina, OH: AI Root Co.
Francis, L.W., Ratnieks, P., Visscher, K. and Vetter, R. (1995) 'Treating AFB using gamma radiation: a case history', *American Bee Journal*, 135: 557–61.
Furgala, B. *et al*. (1989) 'Some effects of the honey bee tracheal mite (*Acarapis woodii*) on non-migratory honey bee colonies in East Central Minnesota', *American Bee Journal*, 129: 195–7.
Hensen, H. and Brodsgaard, C.J. (1999) 'American foulbrood: a review of its biology, diagnosis and control,' *Bee World*, 80: 5–23.
Hornitzky, M.A.Z. (1994) 'Commercial use of gamma radiation in the beekeeping industry', *Bee World*, 75: 135–42.
Johansson, T.S.K. and Johansson, M.P. (1970) 'Establishing and using nuclei', *Bee World*, 51: 23–35.
Johansson, T.S.K. and Johansson, M.P. (1978) *Some Important Operations in Bee Management*. International Bee Research Association, Cardiff.
Kumar, N.R. and Kumar, R. (1997) 'Successful management of a laying worker colony', *American Bee Journal*, 137: 647.
MAFF (now FERA) (2000) *Managing Varroa* (leaflet). MAFF: London.

Matheson, A. and Reid, M. (1992) 'Strategies for the prevention and control of America foul brood,' *American Bee Journal*, 132: 6–8.

Moritz, R.F.A., Southwick, E.E. and Harbo, J.B. (1987) 'Genetic analysis of defensive behaviour of honeybee colonies in a field test', *Apidologie*, 18: 27–52.

Schmidt (2001) 'Pheromonal and hive odour attractants for honey bee swarms', *Journal of Apicultural Research*, 40: 3–4.

Van Eaton, C. and Goodwin, M. (2001) *Control of Varroa: A Guide for New Zealand Beekeepers*. MAF New Zealand: Wellington, NZ.

Vida Apicola (1998) 'El Mal Negro.' 87: 30–5 (in Spanish).

All the above are available from IBRA, 18 North Road, Cardiff CF1 3DY, UK.

SECTION 64:
FURTHER READING

In this final section I have listed what I believe to be the classic books on the science and craft of beekeeping. Every beekeeper will have their own favourites, but this list offers some suggestions for further reading. The main problem with books on bees is that the science of beekeeping tends to advance swiftly (reflecting the importance of bees to humankind) and so books soon become out of date. Some of these excellent books, therefore, have to be read with a degree of circumspection. New books come out every week and it is worth combing Amazon for the latest offerings.

The Hive and the Honeybee (1992 ed.). Graham, J.M. Dadant & Sons, Hamilton, IL. Covers just about everything for all grades of beekeeper from novice to scientist. (It is periodically revised, so try to obtain the latest version.)

The ABC & XYZ of Bee Culture (2005). Shimanuki, H., Flottum, K. and Harman, A. (eds). AI Root Co., Medina, OH. Similar to above but in alphabetical order. Again, try to obtain the latest edition.

Apiculture (1994 6th ed.) Jean-Prost, P. Intercept, Andover. A French perspective. Different and interesting. Written as a series of lessons. Somewhat dated but useful to get a different perspective.

A Manual of Beekeeping (1988 ed.). Wedmore, E.B. Bee Books New and Old, Basingstoke. Covers everything, including things you never through of, but very dated, so be careful with some of the information.

Honeybee Pests, Predators and Diseases (1997 3rd ed.). Morse, R. and Flottum, K. Al Root Co., Medina, OH. A very useful book to have as a reference that covers all pests.

Honeybee Pathology (1991 2nd ed.). Bailey, L. and Ball, B.V. Academic Press, London and San Diego, CA.

Honey Identification (1988). Sawyer, R. Cardiff Academic Press, Cardiff. All about palynology and honey. Very interesting.

A Colour Guide to the Pollen Load of the Honeybee (1994). Kirk, W.D.J. International Bee Research Association, Cardiff.

The Biology of the Honeybee (1987). Winston, M.L. Harvard University Press, Boston, MA.

Honeybee Ecology (1985). Seeley, D. Princeton University Press, Princeton, NJ.

Anatomy and Dissection of the Honeybee (1994). Dade, H.A. International Bee Research Association, Cardiff.

Honeybee Anatomy. The British Beekeepers Association. Stoneleigh Park, Warwickshire. A series of transparent illustrations in sectional form through a bee. Excellent.

Honeybee Brood Diseases (n.d.). Hansen, H. Danish State Bee Disease Committee (also available in Spanish).

Bees: Their Vision, Chemical Senses and Language (reissued 1983). Von Frisch, K. Jonathan Cape, London. A classic.

The Dancing Bees (1954). Von Frisch, K. Methuen, London. Another classic bee-science book.

Bees and Mankind (1982). Free, J.B. George Allen & Unwin, London.

The Honeybees of the British Isles (1986). Cooper, B.A. The British Isles Bee Breeders' Association.

Bee-keeping at Buckfast Abbey (1975). Brother Adam. British Bee Publications, Geddington, Northants. This is a classic book about that famous monk who revolutionised bee breeding.

At the Hive Entrance (1985). Storch, H. Brussels, European Apicultural Editions. Tells you how to recognise certain hive symptoms from looking at what is going on at the hive entrance rather than looking in the hive.

Some Important Operations in Bee Management (1978). Johansson, T.S.K. and Johansson, M.P. International Bee Research Association, Cardiff. Practice and science in action.

Honey Farming (1985). Manley, R. Northern Bee Books, Mytholmroyd, West Yorks. The second-best book on beekeeping that I have ever read. Dated now, especially in regard to disease management.

Following the Bloom (1991). Whynott, D. Stackpole Books, Harrisberg, PA. The best book on beekeepers and their bees I have ever read.

The Bee Garden: How to Create or Adapt a Garden That Attracts and Nurtures Bees (2011) Little, M. How to Books, Oxford. Ideal for those wanting to attract pollinators to their garden and help preserve this all important natural function.

Honey by the Ton (revised ed. 1989). Field, O. Northern Bee Books, Mytholmroyd, West Yorks.

Producing Royal Jelly: A Guide for the Commercial and Hobbyist Beekeeper (2006). Van Toor, R.F. Bassdrum Books, Tauranga, NZ. A step-by-step guide to making a profit from royal jelly; an extra dimension to beekeeping and one that many beekeepers are now taking up.

A Practical Manual of Beekeeping (2011) Cramp, D. How To Books, Oxford. An excellent book all round!

Queen Bee: Biology, Rearing and Breeding (2007). Woodward, D. Northern Bee Books, Mytholmroyd, West Yorks. Written by a New Zealand expert on the subject.

SECTION 65:
BEEKEEPING ORGANISATIONS

The following list comprises beekeeping organisations and websites that have a global interest. Organisations tend to come and go but these have proven themselves robust and will be of interest to readers.

APIMONDIA (International Federation of Beekeeping Associations)
exists to promote scientific, ecological, social and economic apicultural development in all countries and the co-operation of beekeepers' associations, scientific bodies and of individuals involved in apiculture worldwide. APIMONDIA also aims to put into practice every initiative that can contribute to improving apiculture practice and to rendering the obtained products profitable. A major objective of APIMONDIA is to facilitate the exchange of information and discussion. This is done by organising congresses, conferences and seminars, where beekeepers, scientist, honey-traders and legislators meet to listen, discuss and learn from one another (**www.beekeeping.com/apimondia**).

International Bee Research Association (IBRA).
IBRA is a not-for-profit organisation which 'aims to increase awareness of the vital role of bees in the environment and encourages the use of bees as wealth creators'. Its website gives information about the association's mission, members, library services and publications, including the

Journal of Apicultural Research, a major peer-reviewed bee science journal; *Bee World*, a popular beekeeping journal; *The Journal of ApiProduct and ApiMedical Science*, an increasingly important branch of medicine; and *Buzz Extra*, a newsletter for members of IBRA. Information is also provided on conferences and other meetings and there is a useful page of links to relevant websites (**www.ibra.org.uk**).

The British Beekeepers Association (BBKA).

The object of the BBKA is to promote and further the craft of beekeeping and to advance the education of the public in the importance of bees in the environment. Being a member of the BBKA gives you these benefits: *BBKA News*; public liability insurance; product liability insurance; bee disease insurance available; free information leaflets to download; members' password-protected area and discussion forum; correspondence courses; examination and assessment programme; telephone information; research support; legal advice; and representation and lobbying of government, the EU and official bodies. Despite its name it covers England only! (**www.britishbee.org.uk**).

The Scottish Beekeepers Association (SBA).

The SBA is the national body that represents Scotland's beekeepers within the UK, throughout Europe and globally. It is a registered charity and is run entirely by volunteers drawn from its membership. The SBA provides a range of services to its members to assist and support their beekeeping. This includes the use of the Moir Library, based in Edinburgh (**www.scottishbeekeepers.org.uk**).

The Ulster Beekeepers Association (UBKA)

The UBKA is the body to which local beekeeping associations in Northern Ireland are affiliated. Its executive comprises its office bearers who are elected annually at congress and its delegates who are elected by its affiliated associations. The UBKA seeks to promote apiculture and to represent the interests of beekeepers in Northern Ireland while restricting its activities to those which would be difficult for the local associations to perform, such as representation to government regarding registration, support for beekeeping, environmental issues, etc. (**www.ubka.org**).

The Welsh Beekeepers' Association.

This operates along the same lines as the British, Scottish and Ulster Beekeepers' Associations (**www.wbka.com**).

Bees for Development

is an information service at the centre of an international network of people and organisations involved with apiculture in developing countries. It aims to provide information to alleviate poverty and to maintain biodiversity. It is financed from donations, sponsorship, journal subscriptions, training courses, project management, consultancies and information services. Bees for Development promotes sustainable beekeeping to support livelihoods and to conserve biodiversity. They aim to assist people living in poor and remote areas of the world and to raise awareness about the value of beekeeping for sustainable development (**www.beesfordevelopment.org**).

Apidologie

is not an association but a major bi-monthly international journal which publishes original research articles and scientific notes concerning bee science and *Apoidea*. *Apidologie* is a peer-reviewed journal devoted to the biology of insects belonging to the superfamily *Apoidea*, the term 'biology' being used in the broader sense. The main topics include: behaviour, ecology, pollination, genetics, physiology, toxicology and pathology (**www.edpsciences.org**).

CAAPE (Centro Andaluz de Apicultura Ecologico).

Based at the University of Cordoba in Spain. This is singled out because it specialises in solutions for organic beekeepers or those trying to limit their use of chemicals.

And finally of interest to beekeepers the world over,

www.beekeeping.com

the international beekeeping virtual gallery which covers just about everything in the beekeeping world. In English, Spanish, French and German.

INDEX

C

D

E

F

Part J

T

U

W

Y